W0115485

# PLANTS WITH PURPOSE

Gideon Lincecum Nature and Environment Series

# PLANTS WITH PURPOSE

## Twenty-Five Ecosystem Multitaskers

### MONIKA MAECKLE

TEXAS A&M UNIVERSITY PRESS  |  COLLEGE STATION

Copyright © 2025 Monika Maeckle
All rights reserved
First edition

∞ This paper meets the requirements of ANSI/NISO Z39.48-1992
(Permanence of Paper).
Binding materials have been chosen for durability.
Manufactured in China through Martin Book Management.

Library of Congress Cataloging-in-Publication Data

Names: Maeckle, Monika, 1956– author
Title: Plants with purpose: twenty-five ecosystem multitaskers / Monika
Maeckle.
Other titles: Gideon Lincecum nature and environment series
http://id.loc.gov/resources/hubs/60c4f11c-bb03-8bf5-0db5-d08536b41aa4
Description: First edition | College Station: Texas A&M University Press,
[2025] | Series: Gideon Lincecum nature and environment series |
Includes bibliographical references and index.
Identifiers: LCCN 2024033340 (print) | LCCN 2024033341 (ebook) | ISBN
9781648432446 | ISBN 9781648432453 ebook
Subjects: LCSH: Sustainable horticulture—Southwestern States | Restoration
ecology—Southwestern States | Plants, Edible—Southwestern States |
Multipurpose plants—Southwestern States | Conservation plants—Southwestern
States | BISAC: GARDENING / Ornamental Plants |
GARDENING / Landscape | LCGFT: Recipes
Classification: LCC S944.5.C67 M34 2025 (print) | LCC S944.5.C67 (ebook)
| DDC 581.6/3—dc23/eng/20240827
LC record available at https://lccn.loc.gov/2024033340
LC ebook record available at https://lccn.loc.gov/2024033341

All photos taken by the author unless otherwise noted.
All illustrations created by Hilary Rochow.

*Book design by Laura Forward Long*

In memory of my parents, Hilde and John Maeckle,
who taught me resilience and gardening.

Every plant has a story. You just have to dig.

# CONTENTS

# PREFACE

This book has roots in my own fascination with gardening, plants, and insects. As the daughter of German immigrants, one of whom applied his heritage as a village farmer to cultivating tomatoes and radishes in our backyard, I inherited an interest in gardening and a pragmatic lens with which to approach it.

As an adult, I earned my Master Gardener certification in 2004 and became extremely fascinated with the monarch butterfly migration. The journey of this most iconic insect of the Americas unfolds each fall at my family's small ranch along the Llano River in the Texas Hill Country, where monarch butterflies roost in the pecan trees along the riverbank.

There, in 2005, I witnessed firsthand the seasonal rhythms and interactions of wildlife and wildflowers—how the swamp milkweeds emerge from the riverbanks in the spring just as the monarchs are moving north from Mexico to lay their first round of eggs in South Texas. Generally, the agarita blooms first, effusing its redolent sweet scent into the air. The bluebonnets, asters, and bee balms come next, followed by dozens of other wildflowers, with butterflies, bees, and birds showing themselves shortly thereafter.

Later in the season, goldenrod and frostweed line the Llano riverbanks, strutting their showy flowers in late August, offering their nectar bounty to bees and southbound monarchs passing through the "Texas Funnel" on their way to their Mexican overwintering roosts in September and October. The seasonal predictability of these flowers and forbs, and the reliable presence of the wildlife dependent on them, always seems as reassuring as it is beautiful.

As I attempted to replicate this natural showcase in my own San Antonio garden, I started researching these plants and learned of their many

Bee gathers pollen from passionflower.

uses and untold stories. All of this together was so much more interesting than what the plants looked like, which typically is the de facto basis for choosing a plant.

I began transplanting many of these Texas natives from the ranch to my downtown garden and as I literally dug deeper, my journey led to the plants' ethnobotanical, culinary, and medicinal uses. Other useful native and desirable nonnative plants fleshed out my landscape.

What were those little fruits on the Turk's cap I had planted under the pecan trees in our backyard? A little research led me to understand that these "Mexican apples" are edible and contain high doses of fiber, vitamin C, and antioxidants. How many of the plants in my yard could be used as seasonings in food or brewed as tea? And which ones gave back to their garden community by providing resources for wildlife or ecosystem health?

As it turns out, quite a few. And who knew all this was happening in my yard?

It can happen in your yard, too. This book can help you choose plants that serve multiple purposes—for you, wildlife, and the ecosystem.

The COVID pandemic of 2020 bolstered my interest and that of many others in taking stock of our self-reliance and the need to gain a deeper understanding of plants. It's empowering to know you can grow your own wild onions, step into your backyard and grab some bee balm or peppergrass to spice up your dinner, or make tea from many of the twenty-five plants listed here.

Just as important, as we continue to lose more and more species and wild land, it's incumbent upon each of us to do what we can to restore our own corners of the planet for the sake of greater ecosystem health. *Plants with Purpose* aims to help you deliver on these goals by assisting you in making wise choices about the plants you invite to occupy your yard and guiding you toward responsible stewardship of the earth.

Can you eat it? Make tea from it? Does it serve wildlife? Does it provide a health benefit? Is it low maintenance? Does it have a curious story? Does it provide a unique service to the landscape? Can it take the heat? Oh, and is it attractive? These are the questions I asked about each of the plants included here. To qualify, the plant had to answer *yes* to at least three besides beauty.

A caveat: I am not a botanist or a trained horticulturist, but I am a committed and curious student of nature. This book is meant to assist you in plant selection by sharing the amazing stories of twenty-five special plants, and to provide general guidance on how to grow them. More than any other consideration, however, local conditions should determine which types of plants you choose.

I live in plant hardiness zone 9a, for example, which covers much of San Antonio and South Texas. In my zone, a variety of soils and conditions exist. I happen to live near the San Antonio River, so we have pretty decent, loamy soil, but it is highly alkaline. These facts determine what kinds of plants I choose and which ones will thrive in my particular ecosystem. (See the last chapter in this book, "From the Ground Up: Guidance on Planting with Purpose.")

Many excellent resources exist for learning how to propagate plants, amend soil, and garden successfully in your region. I've included some favorites in the resources section. The goal of *Plants with Purpose* is to help you develop your own gardening ethic that works with nature rather than against it by fostering a deeper understanding and appreciation of each plant's particular story. I hope you'll appreciate these plants enough to select, grow, and utilize them.

If your nursery doesn't carry a specific plant you are seeking, ask them to stock it. Growers respond to public demand, so don't be shy about declaring yours. We must work together to create demand for these plants with purpose or they will never be widely available.

# PLANTS WITH PURPOSE

# INTRODUCTION

*Plants with Purpose:*
*Multitasking Plants That Thrive in Warm Climates*

You have space in your garden and are pondering what to plant. Something attractive, for sure—greenery and flowers, some edibles, perhaps. A forb or bush that maybe serves wildlife and attracts butterflies and other pollinators. Space is at a premium—what should you plant?

Plants with purpose.

These days, plants, like the rest of us, must earn their keep. For centuries, Western cultures have prioritized beauty in the garden. Any nursery manager will tell you that shoppers seek "color, color, color." The unfortunate reality is that plant choices by gardeners are largely determined by looks.

Bee snuggles up in a Turk's cap flower.

1

But beauty alone is no longer reason enough to choose a plant that will occupy valuable real estate in your garden or yard. The ecosystem deserves more, and the plants we choose matter. And we owe functional plants—plants that do more than look good—to the ecosystem.

As Doug Tallamy writes in his best-selling book *Nature's Best Hope*, gardening is a lot like cooking. Imagine if you cooked only for taste—what if your menu choices were driven only by tasty foods loaded with fat, sugar, and salt, with no regard for health? Such a single-minded focus would prove deadly in the long run. The same is true for your garden. It's tempting to garden for beauty alone, with no regard for ecological consequences. But such a limited focus can result in a landscape so low in ecological function that it literally drains life from the ecosystem.[1]

Instead, and to carry the cooking metaphor further, consider choosing plants for your landscape as if you were assembling a guest list for a dinner party. Would you choose to spend hours planning, shopping for, and preparing a meal, and cleaning up for a group of guests *just* because they are beautiful? Or would you prefer to assemble and spend time with a diverse group of individuals who are attractive *and* interesting folks—people who offer some type of service to the community at large, a special skill set or expertise, or interesting stories?

When you invite a plant to take up residence in your garden, which is what you do when you make a selection from the local nursery, transplant a volunteer from the wild, or start seeds at home, be advised that you'll be feeding, maintaining, and engaging with this plant for months, possibly years. You'll be investing time tending and harvesting, and using resources like time, water, and perhaps slow-release fertilizer. You'll be living with these plants. Given your commitment—not an unsubstantial investment even with low-maintenance plants like the ones described in this book—shouldn't the plant provide something in return besides beauty?

Such is the thinking behind *Plants with Purpose*. Beauty is a given. Each species introduced and profiled here flaunts its own brand of attractiveness, some more than others. But these plants also serve us and our overextended ecosystem in multiple ways. Each supplies food, utility for those of us who tend it, fuel for wildlife pollinators that make one of every three bites of our food possible, food and shelter for wildlife, or a special landscaping virtue, such as protection (think thorns or ground cover). Recommended plants are also easy to care for, and low maintenance.

Equally important, every plant has a story—some more interesting than others. If you know that the white waxy substance on a prickly pear cactus is the source of a magenta dye used to paint the colonial missions of San Antonio, that Jimson weed played a role in the American Revolution, or that flowering red yucca is actually a type of asparagus, you know fun facts worth sharing. Such awareness makes you and your garden more interesting.

Learning the names and backgrounds of the garden guests with whom we share our landscapes also underscores our tendency to appreciate and protect what we know and understand. Knowing the plant's name is the first step toward that.

Where appropriate, the plant's history and ethnobotanical background is included in its profile, as well as culinary, medicinal, and other uses, and its service to the ecosystem.

I'm not a purist about native plants, but most of those included here are native or desirable nonnatives. Why does that matter?

Our local insects and other wildlife have symbiotic relationships with the particular plants with which they have evolved over millennia. When we import plants that are alien to local landscapes, we're often displacing local flora and undermining or destroying critical interdependencies that make for a functioning planet. Only so many seats exist at the garden dinner table.

In addition, overuse of nonnative or ill-adapted plants can result in a type of horticultural gentrification. Imported species can sometimes exhibit unfair advantages in their newfound contexts, often because they have no natural predators or diseases in their novel environments. As a result, they often thrive and displace native vegetation by hogging available water, nutrients, light, and space. Ultimately, they can oust the locals and create undesirable monocultures, which again can undermine the ecosystem relationships vital to a functioning planet.

Overbreeding and hyperhybridizing is yet another unfortunate consequence of our fixation on beauty. Plant cultivation in a quest for good looks has pushed many of nature's indigenous ecosystem residents out of the commercial nursery market. One of the most common gripes of native plant advocates is that sought-after species are not commercially available. And yet thousands of cultivars, or cultivated varieties, and hybrids are introduced each year. A cultivar is the result of human intervention—that is, plant breeders have inserted themselves and selected plants for specific traits.

For example, dozens of sunflower varieties have been cultivated to bloom without producing any pollen. Some might view these as genetically modified "Frankenflowers,"[2] but florists and wedding planners who make much of their living from brokering the sale of cut flowers told plant growers that they were not pleased with the messy yellow pollen granules that some sunflowers spilled onto furniture and bridal gowns.

The growers responded, and soon they began breeding hybrids of male-sterile sunflowers that produce zero pollen. Dozens of new sunflower cultivars with names like 'Buttercream' and 'Santa Fe Sunset' were born. They usually boast a full-color plastic label announcing their introduction and are frequently found in big-box stores and commercial nurseries.

It's no secret that business and consumer demand drive the garden and nursery market—and that's what happened with pollen-free sunflowers. Yes, they are striking, unusual, and beautiful. But despite being a priority for those in the cut-flower business, pollenless sunflowers are not the best choice for feeding the bees in your yard or helping restore the ecosystem. When growers breed for certain traits or genes, something is often lost. These plants are called "ornamentals" with good reason.

Cultivars of purple coneflower have been manipulated to result in sterile flowers that do not serve pollinators. This native version does.

For example, purple coneflower, a member of the aster family and a coneflower cousin of Mexican hat, has been bred into scores of cultivars with a wide range of colors and double blooms. Such cultivars are often sterile and offer no benefit to pollinators. Their modified flower structures are so complex that "a pollinator couldn't find its way to the center with a map, a compass, and a native guide," according to the Xerces Society of Invertebrate Conservation.[3] And if the cultivar has double blooms, the hybridized blossoms can prevent pollinators from accessing the pollen or nectar altogether. They are almost always sterile. How many of your garden plants inhibit pollination?

Not all cultivars trade ecosystem services for selected traits, however. Nursery sources point out that cultivars like 'Old Mexico' prickly pear, 'Henry Duelberg' sage, and 'Jacob Cline' bee balm have been selectively bred for disease resistance and other desirable traits that don't necessarily harm the ecosystem. Still, the best bet is to choose the unadulterated, unmodified species whenever possible.

In reading these plant profiles, my hope is that you gain an understanding of our complex ecosystems and the inherent connections we take for granted and often disregard, so that you become a better steward in the process. Yes, your tomato plant or Jimson weed attracts a large, green, horned caterpillar that consumes some of the plant's leaves. That's a good thing.

And why would we want caterpillars on our plants?

Because caterpillars serve as the primary food for birds, especially bird parents feeding their young. And when those caterpillars morph into beautiful moths and butterflies, they pollinate our plants, making much of our food and drink possible. They also feed bats and more birds.

**A word about Latin and common names**
The common name, Latin species name, and family name of each plant are included here for the sake of clarity. This is no uppity habit. Taxonomy is famously complicated and ever changing, especially since the advent of DNA mapping. Common names are extremely arbitrary and diverse, depending on the region, so take a moment to research the appropriate species and variety for your situation. For this book, I have used the Lady Bird Johnson Wildflower Center and the USDA Plants Database as my primary sources for taxonomy, but sometimes they

disagree. For ethnobotany and the historical use of plants, I've used the Native American Ethnobotany Database as a primary resource.

Which species you choose to plant in your yard, garden, or container matters.

In general, I've recommended plant families as well as some particular species. A plant family is a group of plants that share similar characteristics—not all that different from human families. They carry similar roots, structures, patterns, and appearances, for example. They also likely share similar growth traits, appearances, nutritional needs, and resistance or receptivity to pests.

Members of the mint family, Lamiaceae, for example, typically exhibit four-sided, square stems and are aromatic. This is useful, because if you want to grow a mint, you will want to look for something in the Lamiaceae family that is adapted to your specific location and ecoregion. The only way to definitively select the right plant for your landscape is to know the plant family and find a particular variety that matches well with your growing region and circumstances. (See the last chapter, "From the Ground Up: Guidance on Planting with Purpose.")

Also, the plant families suggested here typically offer dozens, sometimes hundreds of choices of particular species. It's imperative to assess your garden plot—soil type, sun exposure—as well as grow zone and ecoregion. A plant most likely exists that will meet your needs, but you must do a bit of research to matchmake the species or variety most appropriate for your situation. This has never been easier thanks to online resources and easy access to your local agricultural extension service. You can also quiz knowledgeable nursery or botanical garden staff in your area.

### A word about ethical foraging

When you harvest plants from the wild, please do so responsibly. That means leaving plenty of plants in place so the plant can continue to thrive. Remember, if you take the root, the plant is gone; if you harvest the flower, the plant will not produce seeds. Always leave 90 percent or more of the plant alone when foraging. Use a clean, sharp knife or pruning shears so you inflict the least amount of disease-inviting damage possible, and always obtain permission from the landowner. On public or private property, be sure the specimen is free from pesticides, pollutants, and other contaminants.

And speaking of edible plants, trust that each recipe included here has been kitchen tested by yours truly. Edibility can be a primary or secondary reason for choosing a plant for your yard.

That said, be forewarned that foraging and eating wild food can be dangerous. Be absolutely positive of a plant's identification before you consume it. Also be careful to ascertain that the plant matter is unadulterated and has not been tainted by chemical pesticides or herbicides, and that it has not succumbed to an unhealthy natural disaster, such as flooding by sewer water.

## CAUTION!

Keep in mind when eating herbs or plant parts from the landscape that plant components, even in the same species or family, can vary widely depending on soil, sun exposure, moisture, and locale. Also, personal body chemistry can cause reactions different from those described here. Allergies can also play a role, so tread carefully when trying a new herb or plant you have never sampled.

We learn early on not to run up to wild animals because they could get scared and hurt us. Plants, on the other hand, are rarely given this respect. Unlike animals, plants are rooted in place and are unable to evade predators, which is why they have developed defense mechanisms to protect themselves when a threat looms.

In order to deter predators from eating them, plants might be foul tasting or poisonous like Jimson weed, prick whatever brushes them, as prickly pear does, or release tannins as oaks do, which can cause stomach upset. And keep in mind, just because one part of a plant is edible, that doesn't mean that all parts of it are. Always do your research before consuming any wild food and be aware that plants, just like wild animals, are prepared to defend themselves.

# AGARITA

*Mahonia trifoliolata*

Berberidaceae / Barberry family
Culinary / Defensive landscape plant / Medicinal / Wildlife

Stands of the evergreen bush *Mahonia trifoliolata*, known as agarita, form thickets in the Texas Hill Country and throughout Texas, New Mexico, and Arizona in early spring. Clusters of fragrant yellow flowers appear in February. Red berries follow in April and continue ripening into June.

The thorny bush makes itself known with its prickly leaves and redolent yellow flower clusters that draw bees and butterflies. Songbirds eat its fruits, and quail and small mammals use the thorny shrub for cover.

Myriad creatures favor agarita.

People like Mike Casey, known as the "unofficial mayor of Southtown," a historic downtown neighborhood in San Antonio, Texas, succumb to "agarita fever." This seasonal affliction manifests itself in irresistible get-togethers that include collecting the bright red berries from holly-like agarita shrubs, followed by ritual jelly making.

Casey has been organizing agarita berry harvests from the hedge of his bungalow for more than twenty years. He relishes the tart fruits, sometimes called wild currants, tucked into the prickly shrub.

"I just get in a Zen mood like I'm counting beads," said Casey, who collects the sweet-and-sour fruits bare-handed. "The leaves themselves are firm and pointed like little needles. It doesn't reach the threshold of pain, but it is discomfort."

Agarita berries are the fruit of the most common species of barberry found in Texas. The stickery bush is among the first to bloom in spring. Its fragrant yellow flowers serve as one of the earliest available nectar sources for bees and butterflies. And honey made from agarita pollen has a reputation as some of the most delicious around.

Interestingly, agarita flowers have touch-sensitive stamens, a botanical feature known as seismonasty. Just like a Venus flytrap, an agarita flower will not ignore being touched.

Thorny agarita provides protective shelter as well as food for wildlife.

Juniper hairstreak butterfly nectaring on agarita blossoms.

When a visiting bee or other insect enters the flower in search of nectar, the stamens flail, showering the insect with pollen that can be dropped on the pistils of its next flower visitor.[1] Plus, agarita flowers are entirely edible.

Wildlife also consumes the berries, which are loaded with pectin, carbohydrates, vitamin C, and antioxidants. Agarita seems to be a favorite of many beetles and bugs, as well as preferred fodder for birds, raccoons, opossums, and other wildlife. Deer don't touch it, probably because they can't get to the fruit without being stabbed.

The plant's thorny, five-pronged leaves also rank it high on the most ornery shrubs list. Some folks consider agarita a defensive landscaping tool, planting it in front of windows or fences to keep out intruders. Others use it as a natural fence to deter deer.

Unfortunately, agarita is not widely available in nurseries and it's not that easy to grow. If you find yourself understandably smitten with this plant, you'll have to seek out native plant sales, shop at specialty nurseries, grow it from seed, or dig it up and transplant it from the wild.

If you choose to transplant agarita from the wild, experts recommend digging up a small plant and putting it in a pot for up to a year to stabilize it. You can then relocate it to your garden or yard, preferably somewhere with excellent drainage and lots of sun.

According to *How to Grow Native Plants of Texas and the Southwest*, agarita grown from seed takes three years to produce a five-gallon plant.

Fruit will follow, but the timing is unpredictable and depends largely on light, temperature, and water. Well-drained soil is a must. Seeds are available online, but be sure to get *Mahonia trifoliolata*.

Agarita insists on good drainage and sun exposure and will thrive in a variety of soils, from well-drained loam, clay, and caliche to thin limestone soils. The bush is native to rocky slopes, cliffs, and hills, from coastal South Texas northwest to the Trans-Pecos.

> **Plant type: Perennial shrub.**
> **Light: Full to partial sun.**
> **Water: Extremely drought tolerant.**
> **Soil: Well drained.**
> **Size: 6 feet tall, 6 feet wide.**
> **Bloom color: Yellow, very fragrant.**
> **Fruit: Small, edible red berries.**
> **Bloom time: Early spring, sometimes February, depending on conditions.**
> **Availability: Grab it where you can at native plant sales. If you can dig one up in the wild, keep it in a pot for one year before transplanting to your landscape.**

Some speculate that the plant's name hails from the Spanish word *agria*, which means "sour." Others suggest it comes from the Spanish word *agarrar*, which means "to grab."

Foragers find many uses for agarita. Its plentiful seeds can be roasted and ground for a caffeine-free coffee substitute. Other ethnobotanical sources praise agarita's myriad uses: the leaves can be chewed to prevent nausea, and the roots are utilized to fight fungi and bacterial infections. One application has agarita roots ground up and tucked into sneakers to alleviate athlete's foot.[2]

Agarita's interesting orange wood has been carved into beads and boiled to make dyes. According to seventeenth-century botanist, herbalist, and astrologer Nicholas Culpeper, hair washed with barberry ash lye and water would turn hair "barberry blonde."[3]

Foragers use agarita berries in tarts, pancakes, and cobblers or convert them into jelly, syrup, juice, and shrubs (as in drinks—see below). Some would argue that the toughest part about loving agarita is getting the berries off the bush.

The classic harvesting technique is to lay a sheet on the ground and hit the bush with a broom. A more efficient method is to place an umbrella under a branch flush with berries and comb the spindly branches from the inside of the bush out, since the prickly leaves face outward. Wear protective, elbow-length gloves.

Inevitably, twigs, agarita leaves, dirt, organic matter, and numerous insects will accompany your berry harvest. Once you have your bowl of berries, pick out the debris. Piling a cup or so of berries on a small mesh screen and tilting and shaking it slightly can also help separate the fruit from unwanted debris. Casey dumps the berries on a sheet in front of a floor fan. Our family has developed a method of putting berries on a mesh screen held over a propped-up leaf blower.

### How to make a shrub—the drink, not the plant

You're probably familiar with jellies and jams. Making them typically involves cooking fruit with sugar and often adding pectin, which is a naturally occurring gelatinous starch contained in many fruits and vegetables. Think okra. Depending on the fruit—and what you're trying to make from it—extra pectin is added to boost the jell factor.

I've participated in jelly-making rituals and have enjoyed the results, but I'm not a huge fan of the time-consuming process, especially the sterilization of containers that's required. A negative experience with a friend's blackberry marmalade gone bad literally left a foul taste in my mouth. A mold had crept into the jar, presumably because of faulty procedures. I don't need that kind of pressure and I'm usually in a hurry, which is why I prefer other options for using abundant fruit.

In recent years, I've become a fan of shrubs. What's a shrub? Yes, the word typically describes a short, multistemmed woody plant. But it also refers to a drink laced with vinegar. And currently, the artisanal cocktail circuit has embraced shrubs as a worthy addition to the beverage menu. The word comes from the Arabic *sharab*, which means "to drink." Not surprisingly, it evolved into the more convenient term "shrub." Shrubs require very little time to assemble. My kind of recipe.

Too much fruit and not enough time? Make a shrub!

Shrubs combine fruits, vegetables, and herbs with sugar and vinegar to make a tart, zesty addition to iced tea, a cocktail, or fizzy water. Since a shrub contains vinegar, it is also slightly more healthy, as the fermented liquid helps digestion and your gut biome.

The other advantage of shrubs: making use of extra fruit or vegetables you have on hand. For example, if you have too many blueberries in the fridge or some peaches that need to be used, consider making them into a shrub. Your agarita or cucamelon harvest is out of control? Make a shrub. Mix and match with other fruits, vegetables, and herbs.

## FRUIT SHRUB

*Ingredients*

1 cup agarita berries, American beautyberries, prickly pear fruits, and/or store-bought fruit

1 cup sugar

1 cup white wine vinegar

*Directions*

1. Mix fruit and sugar in a nonreactive bowl (glass or stainless steel is fine). If using larger fruits like peaches, cut them into small pieces and remove pits or seeds and stems. Mash fruit with spoon or potato masher to bring out its juices.

2. Add vinegar and chill for at least 24 hours. The longer it ferments, the more interesting the taste, as it will take on more character if it

sits for a few days. When ready to use, stir thoroughly, strain with a fine sieve, and press the solids against the sieve screen to extract the most juice.

3. Pour shrub mixture into a clean, closeable jar, or funnel into a glass bottle that can be closed with a stopper or cork.

4. To enjoy, add a dash (or more, depending on your taste) to a glass of seltzer, beer, or tea and serve over ice. Or, add a dash to your favorite cocktail. I've been known to use a jigger of shrub in place of simple syrup in my Old-Fashioned, a yummy classic bourbon cocktail that I enjoy during cold months.

*Note: you can add other flavors to the shrub, such as edible flowers, mint leaves, cucumber, basil. Don't be afraid to experiment.*

## MIKE CASEY'S FAMOUS AGARITA JELLY RECIPE
Adapted from Irma S. Rombauer's *The Joy of Cooking*

*Directions*

1. Put 2 quarts of rinsed berries in a pot with enough water to reach the top layer of berries.

2. Bring to a boil and reduce heat to a simmer and cook until the berries split and lose their color.

3. Strain through a jelly bag. Fine mesh bags are available from a paint store.

4. Measure 6 cups of juice and heat to a boil, reduce heat, add 4 cups of sugar, and bring the temperature to 8 degrees above boiling point. (Temperatures for boiling change, depending on elevation.) The bubbles will get very small and exuberant.

5. Pour into sterilized jelly jars.

## AGARITA TART

*My mother always made this frangipane galette with canned sour cherries. It reminded her of her homeland in Germany, where sour cherries are quite common. Over the years, I adapted this recipe into a tart and have used various fruits. One spring we tried it with agarita berries, and it was amazing. The tart, seed-riddled berries add a bit of crunch and fiber to the sweet almond paste base and crust. You can use your favorite pastry or galette dough. I make a basic pastry dough in the food processor. Serve with a dollop of crème fraîche or whipped cream.*

Agarita tart.

*Ingredients*

    1½ cups clean agarita berries

    1 cup frangipane (recipe follows)

    Pastry dough (recipe follows)

    2 tablespoons melted butter

    3 tablespoons sugar

*For pastry dough:*

    1½ cups all-purpose flour

    1 stick, less 1 tablespoon very cold butter (7 tablespoons)

    1 tablespoon shortening

    4–6 tablespoons water

    1–2 teaspoons vinegar or lemon juice

*Directions*

    1. Position oven rack in middle of oven and preheat to 375°.

2. To make the food-processor pastry, use the steel blade of the processor and combine flour, shortening, and butter (cut in chunks) in the bowl. Pulse on and off until you hve a crumbly meal.

3. Slowly dribble spoonfuls of ice water and lemon juice or vinegar into the processor while pulsing. When mixture starts to form a ball, stop processing.

4. Press dough into a flat, round disk, wrap in waxed paper, and refrigerate for 1 to 2 hours.

5. On a lightly floured surface, roll the dough into a circle that will fit your tart pan, with about an inch of overlap for the crust. Lay on tart pan and press edges into side of pan, trimming off excess.

6. Place waxed paper over raw tart dough, then put baking stones, rice, or beans on top to prevent bubbles.

7. Prebake for 10–15 minutes. Remove from oven.

*For the frangipane:*
   3 ounces almond paste
   3 tablespoons butter
   1 egg
   1½ teaspoons sugar
   1½ tablespoons flour

*Directions*

1. In a mixing bowl, beat together the almond paste, butter, sugar, and flour.

2. Beat in egg until completely mixed.

3. Use immediately or store in fridge for up to 2 weeks. Bring to room temperature before using.

*Directions*

1. To assemble the tart, smear the frangipane over the dough.

2. Arrange the agarita berries evenly over the frangipane.

3. Place on middle rack in oven and bake for 20–25 minutes. If dough forms a bubble, pop with a toothpick to let air out.

4. Remove from oven and cool on wire rack.

5. Serve with crème fraîche or whipped cream.

# AMERICAN BEAUTYBERRY

*Callicarpa americana*

Verbenaceae / Verbena family
Culinary / Medicinal / Insect repellent / Wildlife

American beautyberry lives up to its name as a low-maintenance, free-form shrub that's known for clusters of fuchsia-colored berries. *Callicarpa* means "beautiful fruit" in Greek. While bundles of small pinkish flowers grace this medium-sized bush in the spring, bright magenta berries—technically, drupes—later steal the show for great fall color.

Contrary to popular belief, the colorful fruits of this deciduous native bush are nontoxic. Squirrels, raccoons, opossums, and more than forty species of birds feast on the high-moisture berries, which are about the size of a plump peppercorn. The berries have a long life on the plant—

American beautyberries form bundles of drupes.

well into winter when little natural bird fodder is available. Mockingbirds in particular seem to appreciate it as a winter snack. The berries don't have a huge amount of taste off the bush, but some folks swear beautyberry jelly is exceptional, and the fruits can also be made into a tea.

Petite, cuplike pinkish-white flowers emerge in clusters called cymes early in the spring from the leaf nodes along the stems. The fruit follows, and its growth pattern around the nodes suggests the appearance of a brace-let. Bees and butterflies favor the understated flowers early in the season when few flowers are yet blooming. The foliage is host to the rustic sphinx moth, *Manduca rustica*, a brown and gray, white-speckled moth found in much of the United States. Deer forage on the leaves; humans, not so much.

With its arching and ascending branches, American beautyberry works well along a fence line to create privacy, or as a background or understory plant. It can also work as a focal point with its sprays of small flowers in spring and long-lasting show of berries in autumn. As the plant ages, its light green, serrated leaves turn yellow.

The berry stalks make for a lovely addition to a flower arrangement or when plating food.

American beautyberry early blossoms.

American beautyberries start out green.

Mockingbirds love American beautyberries.

**Plant type: Understory bush.**
**Light: Sun to part shade.**
**Water: Drought tolerant but appreciates a regular soaking.**
**Soil: Well drained.**
**Size: 5–9 feet tall, can spread to 5 feet wide.**
**Bloom color: Small pink blossoms in early spring.**
**Fruit: Green berries turn purplish and edible in late summer, fall.**
**Bloom time: April.**
**Availability: Available in nurseries as potted plants.**

I have sampled beautyberries repeatedly and don't find them especially tasty. Nor do I consider tea made from the berries appealing. This is likely just personal taste, given that many find beautyberries delectable. My friend Drake White, owner of the Nectar Bar native plant nursery in San Antonio, makes American beautyberry jelly annually and describes it as "wonderfully aromatic," with "notes of floral ecstasy."

The leaves of American beautyberry have a special chemical constitution that has served humanity medicinally and practically for centuries. The Alabama, Choctaw, Koasati, Seminole, and other Native American tribes relied on American beautyberry for various medicinal purposes. Leaves and other plant parts were boiled for use in sweat baths to treat malarial fevers and rheumatism. The boiled roots were utilized to treat dizziness, stomachaches, and urine retention, while bark from the stems and roots was made into concoctions for itchy skin.

According to oral history, the people of northern Mississippi would stuff American beautyberry leaves under the harnesses of mules to keep bugs away. It wasn't long before folks rubbed the foliage on their clothes or tucked leaves into their hatbands to repel mosquitoes and other annoying insects. Some people would even rub the foliage directly on their skin.[1]

This story was passed down from Mississippi farmer John Rives Crumpton to his grandson Charles T. Bryson.[2] Bryson became a botanist for the Southern Weed Science Research Unit of the Agricultural Research Service in Stoneville, Mississippi. In 2006, he and others determined that American beautyberry leaves contained callicarpenal, which turns out to be a more effective insect repellent than DEET in the battle against mosquitoes and ticks. American beautyberry later earned the distinction of also repelling fire ants.

Further research determined that extracts from the leaves of American beautyberry can stall the growth of the bacterium that causes acne. In 2020, scientists discovered that another of American beautyberry's unique compounds, clerodane diterpene, boosts the activity of a certain antibiotic against antibiotic-resistant staph bacteria.[3]

## BEAUTYBERRY NATURAL INSECT REPELLENT
Adapted from Pepper Gardens Nursery

*Ingredients*
 1 cup crushed beautyberry leaves
 4 ounces rubbing alcohol
 Several drops preferred body wash or spray (optional)
 Small spray bottle

*Directions*
 1. Put crushed leaves into a mason jar.
 2. Cover the crushed leaves with rubbing alcohol.
 3. Let stand for 2 days.
 4. Strain the mixture free of leaves.
 5. Add 2 drops of your favorite body wash or spray for scent.
 6. Pour into a mini personal sprayer.
 7. Spritz on exposed skin or clothing; reapply every couple of hours for best results.

*You can also spray on lawn or porch furniture to deter mosquitoes. And feel free to rub American beautyberry leaves on your skin for an ad hoc insect repellent.*

# BEE BALM / BERGAMOT

*Monarda spp.*

Lamiaceae / Mint family
Culinary / Great cut flowers / Medicinal / Wildlife / Deer resistant

Raucous, fluffy flower heads perched atop whorls of decorative bracts aptly describe bee balm, also known as bergamot or horsemint. Its tubular flowers call to mind a lion's mane or stack of ruffles and prove irresistible to bees, butterflies, and hummingbirds.

Classic prairie wildflowers, *Monarda* species belong in the mint family. They're not fussy about soil, do well in sun or partial shade, and generally require little water. While bee balm typically blooms in late

Bumblebee on bee balm.

spring and can reach up to six feet in height, a trim after first bloom and continued watering can encourage the plant to flower until October.[1] In addition to hummingbirds, bee balm species appeal to small songbirds like finches and sparrows. When the flower heads dry out, the birds have access to the thistlelike seeds.

Different *Monarda* species manifest themselves as perennials, biennials, or annuals. In North America, eighteen species occur naturally, ranging in color from red and hot pink to lavender and white. Commercial growers have embraced their eye-catching flowers and cultivated many varieties with intriguing names like 'Judith's Fancy Fuchsia' and 'Purple Rooster.'[2] The dramatic blooms and lengthy stalks make for great cut or dried flowers.

Bee balm and other *Monarda* species are extremely aromatic and exude a lemony, minty scent that repels fleas and other insects, rendering them rabbit and deer resistant. While bee balm is especially attractive to bees, its name comes from the folk tradition of applying a poultice of the leaves to bee stings and other injuries to relieve pain.[3]

My favorite, *Monarda fistulosa*, has lavender to fuchsia flowers and grows all over the United States. This bee balm is widely available in seed

Bee balm dresses up any border.

Swallowtail on bee balm.

form and occasionally as plugs. I transplanted several of these from the wild around a backyard birdbath, and the lance-leaved stems and succeeding flower pompoms seem to double every year. The bees couldn't be happier. Bumblebees, in particular, seem drawn to this plant.

*Monarda didyma*, another readily available native, has leaves that are especially fragrant. This species is sometimes called Oswego tea because the indigenous peoples of upstate New York taught European settlers to use the leaves as a hot beverage during their boycott of British tea in the 1770s.

Both of these species have a tendency toward powdery mildew in moderate climates, but I haven't experienced that in the Texas heat. Powdery mildew was one reason that growers selectively cultivated varieties that resist the disease, which generally appears in late summer.

**Plant type:** Perennial herb.
**Light:** Sun, part shade.
**Water:** Drought tolerant but appreciates a good soaking. If it has too much moisture, it can develop powdery mildew, a common plant fungus.

Soil: Well drained.
Size: 18–24 inches, depending on species.
Bloom color: Purple, white, dark pink.
Bloom time: May to September.
Availability: Plugs, pots, and seeds generally available. Many varieties and cultivars can be found in nurseries and online, so research what's best for your situation.

The chemical constitution of bee balm has made the plant useful for culinary and medicinal purposes. The Cherokee used the leaves as both a poultice for inflammation and a sleep-inducing tea. The Blackfoot made a bee balm infusion to calm down a cough, and they even used the dried flower heads as makeshift spoons to slurp broth and soup. The Chippewa chewed up the leaves and placed them in their nostrils to alleviate headaches. Flatulence, fever, saddle sores, stomach cramps—the ethnobotanical uses of bee balm are vast.

Besides a tasty, calming tea, bee balm's fragrant leaves can be used as a seasoning similar to oregano or mint. As with any fresh edible, the age of the greens upon harvest can influence how they taste.

Kelly Kindscher writes in her book *Edible Wild Plants of the Prairie* that the younger leaves of *Monarda* are best utilized for tea, the later leaves (until the plants flower) make the best oregano-like seasoning, and the old leaves are the hottest and can be used in hot sauce.[4] Flowers make a great edible garnish.

Biologist and forager Jim Meuninck suggests adding a half cup of bee balm flowers to two cups of your favorite meat marinade for a lemony, oregano flavor.[5]

### How to Make Herbal "Tea"

My serious foraging friends point out that technically, a true "tea" can be so called *only* when made from the *Camellia sinensis* tea plant, which is native to Asia. We use the word generically here to refer to a drink brewed from herbs.

Bee balm can grow to two feet tall.

My impatient and pragmatic approach usually involves throwing dried herbs like bee balm, for example, into a pot, bringing them to a boil, and then turning off the heat. I let it steep from five minutes to overnight and then strain the mix (technically an infusion by ethnobotanical standards) into a pitcher for a ready batch of herbal iced tea. For hot tea, the amount of herb and time brewed would be less. You can also do it in the microwave.

For best results, dry the plant matter first. Two weeks is ideal. Drying rids the plant of chlorophyll, which is distasteful to most palates.

Simply gather the leaves, fruits, or flowers you intend to use, rinse off bugs and detritus or dirt, and clip or tie into bunches with a string. Hang the bunches upside down in a dark, well-ventilated space to dry out. They're ready when the leaves crackle. If you're in a hurry, you can dry the plant matter in a low-temperature oven or dehydrator.

Once the leaves are dry, throw them in a food processor to grind them up, or crumble by hand, then store in airtight jars. With flowers or

Once they're dried out, bee balm petals can be used to make tea.

seeds, watch for mildew and don't use. You can use a tea bomb to brew, free float the dried herbs in a pot of water, or even purchase empty tea bags to fill up and use as needed or to give as gifts.

When the leaves are ready, put them in a pot, cover with water, bring to a boil, and turn off the heat.

Keep in mind that the longer a tea brews, the stronger it gets. Sometimes teas can get bitter if you stew them too long, so feel free to experiment. Personal taste will dictate brew time. The stage of the plant (how young or old it is) can also affect the taste. You'll learn this over time.

Some recommend harvesting leaves and flower parts early in the morning on sunny days after the dew has evaporated. This helps avoid dampness, which is generally undesirable (unless you'll be using them immediately), since that can cause mold.

Choose healthy-looking foliage and flowers. If you're snipping a plant from the wild or a public space, make absolutely sure it has not been tarnished with pesticides, auto exhaust, or other possible contaminants.

# CHILE PEQUIN OR CHILTEPIN

*Capsicum annuum*

Solanaceae / Nightshade family
Culinary / Medicinal / Mammal repellent / Wildlife

Chiltepin and chile pequin are two peppers frequently confused with each other. Both are classified as *Capsicum annuum* and both provide a carefree, attractive, and spicy addition to the landscape.

Chiltepin peppers on the bush.

Chiltepin is a wild pepper native to the Amazon rain forest of Bolivia and Brazil. The tiny, ever-blooming white flowers and resulting petite fruits of this mother of all peppers are comparable in size to a large holly berry or bloated peppercorn. *Tepin* comes from the Aztec Nahuatl word that means "flea," and *chiltepin* has been appropriately translated as "hot flea."

The fruits start out green and turn red as they mature, building their capsaicin, a heat-rendering alkaloid, in the process. Birds love them, and people do, too. New Mexico State University's Chile Pepper Institute describes chiltepin as an herbaceous plant "favored by birds and masochists."

Chile pequin is a cultivated variety of the wild ancestor. It has slightly more elongated pods but emulates the growth pattern of a stocky bush—durable, low maintenance, drought tolerant, and überuseful. *Capsicum annuum*, known as chiltepin or chile pequin, also goes by chilipiquin, bird pepper, and turkey pepper. It is one of twenty-seven species in the *Capsicum* genus, five of which have been domesticated. Chile pequin was named the official state native pepper of Texas in 1999, while its more widely known offspring, the jalapeño, preceded it as the official state (nonnative) pepper in 1995.

The flowers of chiltepin are tiny and white.

Most chiltepin peppers are planted by birds.

The plant makes an attractive understory bush in the landscape and requires little care. It can serve as a woodsy backdrop or a potted plant, and the more you pamper it with supplementary water, the more densely it seems to grow. The half-dozen chiltepin bushes in my downtown garden were all planted by birds. They are rarely watered and never fertilized, and yet they thrive. Their tiny white flowers start showing in the spring and continue through fall, quickly developing green fruits that turn into red hot chiles.

Chile peppers, including these two, are among very few plants on the planet to produce capsaicin, the heat-rendering chemical that gives peppers their punch.[1] Interestingly, most capsaicin is stored in the pepper's placenta, the white membrane inside the pepper walls to which the seeds are attached. When the peppers dry, the capsaicin can transfer to the seeds or even the pepper wall, perhaps contributing to the often held misperception that the seeds contain the most heat. In short, if you want to limit heat on a fresh pepper, remove the white placental membrane to which the seeds are attached.

Pepper plants developed capsaicin as a defense mechanism to improve their chances of reproduction. The bright red berries of chiltepin were so attractive to mammals that they would consume the small pods, decimating the seeds through chewing and digestion. The result? The plant's range and reproduction were grossly limited.

Chiltepin adapted by boosting its capsaicin production, making the enticing fruits less appealing to pepper-consuming mammals that have heat sensors in their mouths. Birds, meanwhile, have neither the same heat receptors as humans, nor the teeth or brutal digestive tracts that render seeds infertile. Avian species can eat and pass the seeds, thus "planting" them all over America, which is why we often find wild chiltepin and chile pequin along fence lines and under trees.

Chiltepin found its way north via bird droppings and was later cultivated into the chile pequin, a slightly larger, more oblong pepper. Since *pequin* comes from *pequeño*, which means "small" in Spanish, it makes sense that the fruits and flowers of the chiltepin are petite.

But don't let the small size fool you. The mighty chiltepin packs a relatively hot punch on the pepper heat-ranking system known as the Scoville scale—40,000–60,000 Scoville Heat Units, compared to the jalapeño's 2,500–8,000. The Scoville scale is a ranking system developed by pharmacologist Wilbur Scoville in 1912 as a way to measure the heat or spiciness of peppers. For many years, the spicy pepper rating system has been the only way to measure the heat or spice level of peppers. Under Scoville's supervision, trained pepper tasters sampled different pepper extracts that had been diluted in sugar water in increasing degrees until the heat level could no longer be detected. The heat level was then rated in multiples of one hundred "Scoville Heat Units." For example, the Carolina reaper, one of the hottest peppers on the planet, ranks a searing 2.2 million Scoville Heat Units, about two hundred times hotter than a jalapeño and forty-five times hotter than a chile pequin.

Researchers have since figured out a more scientific way to measure the actual capsaicin content of given varieties, but the Scoville scale continues to be cited and utilized by chileheads competing in pepper-eating contests for bragging rights. One contestant ate so many hot peppers that he burned a hole in his esophagus and had to be hospitalized for weeks.[2] Psychologists have explored the fascination of those who enjoy the pain associated with eating hot chile peppers and determined that such folks are thrill seekers. They compare the sensation of extreme chile heat on the tongue to the satisfaction that accompanies a terrifying roller coaster ride.[3]

   Besides their ease of growth, novelty, beauty, and edibility, chile pequin and chiltepin serve wildlife extremely well. Birds devour the bright red berries. Squirrels, deer, and other mammals leave them alone. If you have a squirrel problem at your bird feeder, add some chile powder.

   Chile pequin are amazingly versatile and easy to grow. As mentioned, birds can process the seeds as food without destroying them and often plant the pepper bushes wherever they poop. The bushes, which can reach four feet in height and an equal width, grow in sun and shade, with or without water, and return annually, even after a deep freeze.

Ready to make salsa?

Plant type: Perennial herb.
Light: Prefers partial sun/partial shade.
Water: Drought tolerant.
Soil: Well drained, alkaline, adaptable.
Size: 4 feet tall, 4 feet wide.
Bloom color: White.
Fruit: Tiny green berrylike fruits follow the flowers, turning red and hotter in taste as they remain on the bush.
Bloom time: Spring to fall.
Availability: Seeds and plugs available online. Nurseries often carry it, especially native plant nurseries. Can also grow from cuttings if you have access to wild plants or a friend's chiles. Cuttings should be 6 to 8 inches long and grown in perlite or organic potting soil. They should root easily if kept warm and misted and should be ready for transplanting in 6 to 8 weeks. Seeds can be started indoors and transplanted outside as appropriate.[4]

Chiltepin and chile pequin offer as much versatility in the kitchen as they do in the landscape. Dry and crush them for use as spices on pizza and in soups, stews, beans, and salsas. Add to mayo for a picante kick, steep them in vinegar for a spicy condiment, or sprinkle on scrambled eggs or chocolate ice cream (yes, it's a thing!).

Chile pequin are free, organic, and delicious. Better to grow your own, given their high price tag. Prices for dried chiltepin peppers range from $12 for a quarter ounce to $100 or more per pound.

Other uses include grinding them and sprinkling the flakes and seeds in your yard to keep dogs, cats, and other critters from digging up recently planted seeds and seedlings. You can even make your own capsaicin arthritis cream from chile powder and coconut oil. Or, string a crop of chile peppers together for a festive holiday wreath or *ristra*—the good luck charm of New Mexico that is kept in the kitchen for culinary use.

Culinary and medicinal uses of peppers have a long history. Chile peppers work as a natural insect repellent, antifungal remedy, arthritis treatment, cold medicine, and preservative. To wean a child, the Navajo rubbed powdered *Capsicum annuum* on the breast of nursing mothers. The child learned an immediate distaste for breast milk.

# CITRUS: ORANGE, LEMON, LIME

*Citrus spp.*

Rutaceae / Rue family
Culinary / Medicinal / Wildlife

Hot climates welcome citrus in many forms—oranges, lemons, kumquats, and limes. The species offers myriad charms: dark, evergreen foliage, extremely fragrant blossoms, colorful, edible fruit, and health benefits.

Citrus bloom in January 2020.

Citrus is also a host plant to the giant swallowtail butterfly. Plant just about anything citrus and you'll have a reason for this magnificent butterfly to visit your yard. The large lepidopteran's range covers most of the United States. It flaunts a wingspan approaching six inches and shows dramatic black and yellow on the dorsal side of the wings, with an underside of yellow, blue, orange, and black when wings are closed. It also sports a distinctive yellow abdomen.

Females lay cream to brown eggs individually on the upper surface of citrus leaves. Caterpillars usually feed at night. In Florida's orange groves, giant swallowtail caterpillars are known as "orange dogs" for their ability to decimate citrus orchards.

Giant swallowtails mimic bird poop very convincingly in their early caterpillar stages. Glossy, brown-and-cream skin sheens like fresh excrement, creating an unappetizing front for predators. When poked or bothered in their later growth stages, they flash their osmeteria, reddish antennae-like extensions imitating the forked tongue of a snake. Their chrysalis, like that of the eastern black swallowtail, includes a silk saddle that holds it upright.

Bird poop? Nope. Giant swallowtail caterpillars on citrus.

Another reason to plant citrus: the plant attracts bees. "Orange Blossom Special" is not just a great Willie Nelson ballad. It describes the unique flavor notes of pollen and nectar harvested from citrus trees. Sought-after orange blossom honey is loaded with antioxidants, minerals, and antibiotic properties—plus, it tastes amazing and is exceptionally aromatic.

ANOTHER member of the Rutaceae family, *Ruta graveolens*, commonly known as rue or herb-of-grace, serves as host plant to the black swallowtail butterfly (see chapter on fennel). Rue is evergreen and easy to grow, has grayish-green foliage, and is widely available. The plant you find in your local nursery is an heirloom with no cultivars or varieties propagated since.[1] It has a distinct smell that some folks find offensive, but I find fragrant.

The plant grows into a small shrub and requires little attention yet draws butterflies in spring and summer looking for a place to lay their eggs.

First-instar swallowtail caterpillar on rue.

Swallowtail caterpillar on rue flaunts its tubercles.

Rue is one of the oldest medicinal herbs and has been used to treat everything from inflammation and ulcers to arthritis, and as an insect repellent or tea. Some people have a strong reaction to tactile contact with rue, so check your sensitivity or wear long gloves until you know how you personally respond to its touch.

Subtropical by nature, citrus can be tricky to grow, depending on your soil type and how hot or cold your location gets. Citrus trees do *not* like freezes. Lime and lemon trees have a reputation as some of the most cold-tender citrus.[2] That said, some species can take a jolt of frigidity better than others.

Consult with your local agricultural extension or independent nursery to see what varieties are best suited to your location. Most citrus transplanted in containers will produce fruit within two to three years when cared for properly. Bigger trees, which can get as tall as twenty feet, produce more fruit.

Be advised that certain orange, lemon, kumquat, and lime varieties can be grown in containers so that the plants can feed butterflies and bees in warm months and then be wheeled inside when a freeze looms. Planting citrus trees in a protected outdoor setting such as a breezeway or an urban yard surrounded by heat-generating concrete or housing can also protect citrus from succumbing to cold. Just keep the trees six to eight feet from any structures to avoid root or canopy issues.

Trifoliate orange, one of many citrus species, produces seed-riddled fruit.

Plant type: Small tree.

Light: Full sun. Keep in mind that all citrus are subtropical and will *not* tolerate a freeze. If you are likely to get a freeze and are not comfortable losing your tree, then plant yours in a pot.

Water: Medium. Water regularly, but don't allow standing water, or water remaining in the dish if you plant your citrus tree in a pot.

Soil: Deep with good drainage.

Size: 10–20 feet tall, depending on species; 15–20-foot spread.

Bloom color: Cream-colored flowers in late spring bear fruit later in summer.

Bloom time: April to November, sometimes later depending on climate.

Fruit: Orange, yellow, green—color and size depend on what you plant.

Availability: Check with your local nursery for starter trees in pots. Most citrus are root grafts, cultivated for specific soil and conditions, so shop for your preferred fruit and form based on what grows well in your area.

Trifoliate orange produces thorny spines and can be used as a defensive plant.

**Note: citrus usually does not produce fruit until its third year, so be patient. Also, if your citrus tree is a root graft and freezes, it will likely grow back from the root but may be sterile and no longer produce fruit.**

The edible and drinkable benefits of this plant family are well known, dating back to ancient times when lemons were only for the rich. The most ancient citrus, known as citron, is believed to have been introduced by Alexander the Great from India to Greece, Turkey, and North Africa in the late fourth century BC.

Flash forward to the seventeenth century, and one could argue that citrus played a key role in conquering the world. Maritime explorers like Vasco da Gama and Captain James Cook battled scurvy—the "plague of the sea." The disease, caused by a vitamin C deficiency, was typified by weakness, loose teeth, bleeding, and the breakdown of collagen and connective tissue. Scurvy often resulted in a painful death, decimating sailing crews and cutting short maritime exploration.

But in 1747, James Lind, who became known as the father of naval medicine, made citrus a must-have on maritime voyages. Lind, the son of a London merchant and medical apprentice, took a dozen men who appeared to be suffering from scurvy, divided them into six pairs, and treated them with various remedies, including cider, herb mixtures, vinegar, and citrus. Those who consumed two oranges and one lemon a day were well enough after a week to help nurse the others.[3] Appreciation for vitamin C, or ascorbic acid, which is found naturally in citrus fruit, continues today.

## MONIKA'S MARGARITAS

*Ingredients*
   1 part fresh-squeezed lime juice
   1 part orange liqueur (*not* triple sec)
   1 part 100% agave tequila
   1 part water

*I usually make a batch of 4 cups of margaritas at a time, which makes about 5 drinks of 6 ounces each. It keeps well in the fridge. However, you can do as little as ¼ cup of each ingredient and make 8 ounces for 1.5 drinks, if you choose.*

*Directions*

1. Squeeze limes. I usually make a cup.
2. Add equal parts orange liqueur (I prefer Citrónge), tequila, water.
3. Salt rim of glass.
4. Fill halfway with ice.
5. Load shaker with ice and fill ¾ full, then *shake shake shake* until you hear the sound of ice retreating into the foam your shaking has produced. When it sounds muffled, it's ready.
6. Strain shaken margarita mix into prepared glasses.
7. Garnish with a slice of lime or citrus flower. *Enjoy.*

# CUCAMELON

## *Melothria scabra*

Cucurbitaceae / Gourd family
Novel / Culinary / Wildlife / Ground cover

If ever a contest were staged to name the most adorable fruit, the cucamelon would likely win. These grape-sized, watermelon-like fruits dangle from delicate climbing vines fastened to screens and fences with intricate curly tendrils. Their mottled green-and-cream pattern suggests watermelon, yet they taste like cucumbers. And the petite orbs exude charm and intrigue, as if they'd fit right in at a tea party for fairies.

Harvested cucamelon.

If you want to get your child hooked on gardening, plant cucamelons. Kids *love* this plant. Like sunflowers, they are easy, fast growers and big producers. The fruits are edible off the vine and ignite imaginations. It's possible that the primary purpose of cucamelons is to get people interested. Inevitably, they raise questions such as "What the heck is this?"

That was my reaction when I first encountered the dainty fruits. For weeks I hadn't noticed the delicate vine climbing up our screened porch, and then suddenly the tiny melons were everywhere. A quick internet search for "tiny watermelon fruit" led me to the conclusion that the uninvited cucamelons were safe to eat. That first bite of the freckled green fruit released a tangy, cucumbery taste—fresh, crunchy, with a slight sour pop. I hadn't planted this volunteer and figured it had migrated to my garden via potted plant, since birds don't seem to consume it. It has returned each spring for the past three years.

Known as mouse melon, Mexican sour gherkin, and *sandita* in Spanish (which means "little watermelon"), cucamelons have been around since the time of the Aztecs and moved north from Mexico and Central America. In recent years they've captured the attention of foodies, farmers' markets, and social media gardening influencers, carving out a reputation as the coolest cucumbers on the planet—except they're technically not cucumbers. They're gourds.

This volunteer cucamelon vine climbed all over my screened porch.

The Cucurbitaceae family contains lots of familiar crops, including cucumbers, cantaloupes, squash, and gourds, but only one species, *Cucumis sativus*, is considered a true cucumber. Cucamelons belong to a separate genus, but *Melothria scabra* has been granted honorary cucumber status because it mocks the real thing so well in both growth habit and taste. Because of their ease of growth and lack of pests, cucamelons serve multiple uses in the landscape. They can form a trellis or climb a screened porch, or you can create a fort or hut with sticks adorned with the rampant vine. It even works as a ground cover—although you might step on the yummy fruits if you don't pay attention. Bees and butterflies enjoy the tiny flowers, helping to pollinate the male and female blooms. Note: the female flowers emerge first.

CAUTION!

Another species, *Melothria pendula*, looks similar to cucamelon and has been highlighted by the popular foraging expert Mark "Merriwether" Vorderbruggen. Uncommon in the wild and known as creeping cucumber, *Melothria pendula* grows in moist areas, turns dark purple when ripe, and is best avoided, except when green. Eating the dark, ripe fruit can result in a powerful laxative consequence, according to Vorderbruggen. "Its bowel-purging effects hit very rapidly and very uncontrollably and can result in serious injury to the body from dehydration."

Cucamelon fruits dangle daintily from the vine.

In my San Antonio yard, cucamelons continue producing well into December, and the more we harvest, the more fruits we get—more than one hundred fruits per vine.

Seeds are readily available online, and once you establish a plant you can save your own seeds or hope for self-reseeding annually, which has been my experience.

If you choose to save seeds, choose the ripest fruits. Take the melons indoors and let them stand a week or two on a tray to ripen further. Then cut them up, remove the mucous membrane, and take out the seeds. Rinse them off and move to a paper towel to dry. Make sure the seeds are dry before storing in a jar, or mold can result. Store in a cool, dry place and plant them the following spring, depending on your climate and ecoregion. Another option: keep the whole fruits in a cool location above freezing and plant the entire cucamelon in the spring.[1]

Cucamelons, like cucumbers, need about six hours of sun a day to fruit. Unlike cucumbers, they don't require much water, which makes them ideal for hot, dry climates. They're so resilient, drought tolerant, and productive that Utah State University included cucamelons in a 2021 paper assembled to promote the cultivation of new kinds of produce suitable for increasingly arid climates.[2]

**Plant type:** Climbing perennial.
**Light:** Full sun.
**Water:** Cucamelons are famously drought tolerant once established. Allow to dry out between waterings.
**Soil:** Well drained.
**Size:** Rampant grower; can climb 20 feet in ideal conditions and spread 3-4 feet in width. Can also work as ground cover.
**Bloom color:** Yellow.
**Fruit:** Tiny green melons.
**Bloom time:** Spring to fall.
**Nuisance potential:** Rampant grower; can climb 20 feet in ideal conditions and spread 3–4 feet in width.
**Availability:** Easy to grow from seeds.

Note: some afternoon shade is beneficial in the heat of summer. While cucamelon is known as a climbing vine, it works as a ground cover if it has nothing to assist its vertical growth. It can also be grown indoors in a pot in a sunny window or in a basket from which its petite fruits will dangle. The more you harvest the fruits, the more productive the plant will be.

Cucamelon flower.

Cucamelon fruit follows the flower.

Cucamelons' edibility is versatile. Eat them straight, toss them in salads, spike them on toothpicks and serve with a cube of cheese, use them as a garnish on cocktails, or puree them to make a cucumber-like juice. Some argue they're tastier marinated or pickled.

References to the Aztecs using cucamelons as sustenance show up in various journals, but cucamelons seem to have been overshadowed by research on the more common and commercially popular cucumber species. The James Beard Foundation cites them as part of the Aztec diet and states that cucamelons are now "commonly served as a delicacy in Central America." Given their cute appeal, ease of growth, and utilitarian capabilities, it's surprising that more farmers are not growing cucamelons.

No mention of cucamelons as a specific medicinal treatment exists in resources like the Native American Ethnobotany Database, perhaps because the plant is native farther south. However, the cucamelon's more potent native cousin, *Melothria pendula*, was cited as a snakebite remedy for the Houma tribe.

Cucamelons offer many of the general health benefits of cucumbers. They're low in calories, high in water and fiber, and loaded with antioxidants, phytonutrients, and even calcium. Health advocates recommend eating cucumbers with skins on and seeds intact because that's where most of the nutrition resides.[3] With cucamelons, that's a given, and the peel has no bitter taste like that of some regular cucumbers.

A 2019 study on cucamelons suggests that its leaves may hold potential as an antidiabetes treatment because certain cucamelon extracts are excellent at absorbing glucose.[4]

Cucamelon
bounty.

# FENNEL

### *Foeniculum vulgare*

Apiaceae / Carrot family
Culinary / Medicinal / Wildlife / Pollinator

Members of the carrot family—parsley, dill, and fennel—work overtime in the yard as herbs for cooking and as a host plant to the common Eastern black swallowtail butterfly. Depending on where you live and the time of year you're planting, one of these plants will serve well.

Fennel blossoms.

In hot climates like those of Texas, parsley can take some heat but often dies back. As a biennial, it will return the following year, but this can leave a hole in your gardenscape. Dill seems to be the most delicate of this trio, and it shrivels as soon as temperatures heat up.

Fennel, on the other hand, while known as a cool-season crop, is an almost year-round superstar, even in warm Texas climates. This native of the Mediterranean has wispy fronds that add a delicate touch and appealing fragrance to garden borders and herb beds.

Two types of fennel are generally available: *Foeniculum vulgare*, which is treated as an herb, and a variety known as Florence fennel, or finocchio (rhymes with Pinocchio), which produces bulbs.[1]

Both of these fennels are entirely edible: seeds, pollen, fronds, stalks, bulbs. Finocchio, however, is the one to plant if you want to harvest fennel bulbs, which can be treated like a vegetable. Look for the *Foeniculum vulgare* variety 'Finocchio' at your local nursery.

Plugs of both are widely available, but you can also start the plant from seed. The bulbs thrive in cool temperatures, and as summer heats up, the delicate fronds push out clusters of tiny yellow flowers that draw numerous insects. Later, tasty seeds form that are prized in Ayurvedic medicine circles for their boost to the digestive system.[2] Bees, ants, wasps, and others gravitate to the thick pollen and sweet nectar.

Swallowtail caterpillar on fennel.

Edible fennel foliage on left, edible dried seeds on right.

Plant type: Perennial herb.
Light: Likes full sun but will die back in hot summers; generally a cooler-weather plant.
Water: Moderate.
Soil: Well-drained, rich loam.
Size: Florence fennel grows to about 2 feet; *Foeniculum vulgare* can get as tall as 6 feet.
Bloom color: Yellow.
Bloom time: Late spring, early summer.
Availability: Easy to find in the herb sections of nurseries. Can also be grown from seed.

In recent years, fancy chefs have started using fennel pollen by dusting it on pasta, salads, and other edibles. With its hints of licorice, citrus, and honey, fennel pollen has earned a reputation as a magic ingredient and is known in Tuscany as the "spice of the angels." Fennel seeds can be brewed as a medicinal tea that relieves flatulence, or they can be toasted and sprinkled on salads, baked goods, roasted veggies, or meats.

Harvesting the seeds is easy. As summer heats up, the yellow fennel flower heads will dry up and produce flat seeds that taste like licorice. When you notice the seeds drying out and dropping, grab a bowl or paper bag and shake the fennel plant to capture the seeds. You can also cut off the flower heads and bundle them with a twist tie or rubber band, heads down, to dry out completely. Leave them suspended upside down in the bag for about two weeks and then shake the heads so the seeds fall to the bottom. Once they're completely dry, store in a glass jar and use as needed. Make sure no moisture remains—it will cause them to mold. Tip: the seeds are much tastier toasted than raw.

You can eat fennel raw and sliced, tossed with oranges or apples and dressed like a salad. Braised in the oven or a pan, or grilled, it complements Italian sausage as a main course. Note that if you're planning to harvest fennel bulbs, you'll want to do so early in the season and not allow the plant to bolt, or go to seed. This will make the bulb less sweet, as stored sugars will be spent on flower production rather than retained in the root.

## SAUTÉED FINOCCHIO SALAD WITH LEMON–FENNEL SEED DRESSING

*Ingredients*
   ⅔ cup olive oil
   ⅓ cup fresh lemon juice (about 2 lemons)
   1 tablespoon fennel seeds, toasted
   Kosher salt, freshly ground pepper
   Fennel bulb
   Greens of your choice
   Parmesan cheese (optional)

*Directions*
   1. Rinse fennel bulb and/or stems. Pat dry. Cut in quarter-inch slices, set aside.

2. Heat sauté pan with no oil, and toast the fennel seeds about 5 minutes, until fragrant. Set aside.

3. Heat 1 tablespoon of oil in sauté pan, then add sliced fennel so that it makes a sizzling sound. Cook on medium heat. Stir as needed to prevent sticking and burning, about 10 minutes.

4. While fennel is cooking, assemble greens of your choice—arugula, romaine, butter lettuce, frisée. Set aside.

5. Once fennel is cooked, remove from heat.

6. Mix lemon juice with olive oil, salt, pepper, and toasted fennel seeds.

7. Dress greens, add fennel last. Sprinkle with grated Parmesan cheese.

*Note: toasted fennel seeds make a delicious and healthy addition to salads, muffins, or garlic bread.*

WE'RE NOT the only ones who like to eat fennel. The Eastern black swallowtail, *Papilio polyxenes asterius*, consumes fennel foliage as a caterpillar.

Eastern black swallowtails can be found in the garden throughout the growing season. They rank as one of the most prolific butterflies in the United States. Males and females differ slightly in appearance, with both showing dramatic black outer wings and double rows of yellow spots—larger and brighter in the males, more understated in the females. A swish of showy powder blue graces both sexes but is more pronounced in the females.

Swallowtail egg on fennel.

Swallowtails will find your parsley, dill, or fennel and lay round, gold eggs on it. The caterpillars change appearance dramatically from the first instar to the fifth, starting out as a dab of orange and black with a white-gray "saddlebag" in the middle, before morphing into a dramatic black, yellow, and green caterpillar, sometimes confused with a monarch.

One amusing aspect of swallowtails is their "stink horns"— bright yellow tentacle-like "tubercles" that reveal themselves when disturbed. Upon the mildest poke or prodding, the swallowtail caterpillar juts out a forked yellow gland called the osmeterium to show that it perceives danger. The tubercles emit a distinctive sickly sweet odor, presumably as a warning to predators. Kids are always impressed when you provoke a swallowtail's tubercles.

Swallowtails are famous for wandering far from the host plant and taking their time to emerge from the chrysalis at unpredictable moments—from weeks to many months. Chrysalises formed late in the season often overwinter until spring. Their unpredictability is also manifested in the varied color of the chrysalis that results from the final morphing. Sometimes brown, sometimes green—you just never know what color a swallowtail chrysalis will be.

When a swallowtail makes its chrysalis, it will bow its head in an upside-down J shape and spin a silk button to attach itself to a twig, branch, or other surface. It also forms a silk harness to hold the chrysalis snugly in place during the time it takes to transform its DNA into a butterfly—again, an often unpredictable span.

Swallowtail caterpillar shows off its tubercles.

Swallowtail chrysalises in many colors.

# FROGFRUIT

*Phyla nodiflora*

Verbenaceae / Verbena family
Culinary / Medicinal / Excellent alternative ground cover / Wildlife

If you're looking to replace nonproductive lawn with a plant that provides a major benefit to wildlife, frogfruit is an excellent bet.

This versatile ground cover can take wet feet or long dry spells and does well in full sun or semishade. It maintains a tidy height of three to six inches and holds up well to foot traffic. Frogfruit requires little maintenance—no mowing or fertilizing, and after getting established, little supplemental water. Even in scorching Texas summers, frogfruit continues to produce tiny whitish-pink flowers, drawing bees and butterflies.

Frogfruit poses for its glamour shot.

Native to the southern half of the United States, frogfruit can be gregarious, depending on conditions. If it gets too rambunctious, it can be mowed and will recover. It spreads rapidly from its woody taproot, often creating a mat as it pushes out trailing stems and incessant flowers from late spring through fall. The flowers sit on what appears to be a cylindrical rust-colored cluster. The foliage stays evergreen, except when tested with a frost.

For those who live in hot climates that prioritize water conservation, frogfruit offers a great alternative to water-guzzling grass lawns.

The plant is irresistible to bees and butterflies as a nectar source. The lance-shaped leaves work as host fodder for three different brush-footed butterflies. The Phaon crescent, white peacock, and common buckeye will all lay their eggs on frogfruit.

All three butterflies are beautiful, but the common buckeye's distinctive eyespots set it apart and make it easy to identify. The creature gets its name from the resemblance of its eyespots to the eyes of male deer (bucks). With its wings closed, the buckeye's large, bluish eyespots create the mirage of a threatening owl and help intimidate predators. Its scales of brown, bluish purple, cream, orange, and black paint a lovely profile.

Frogfruit's tiny flowers attract many pollinators.

Even in the brutal heat of a Texas summer, frogfruit makes a versatile, resilient ground cover.

Male buckeyes have a reputation for flying low and perching during the day to keep an eye out for females. Females lay spherical green eggs singly on frogfruit and other members of the Verbenaceae family, generally preferring leaf buds or the upper side of host plant leaves. Caterpillars are black and spiny, with orange accents and occasional hints of purple.

The eggs hatch in seven to ten days, morph through the caterpillar stages in eight to twelve days, and then form a chrysalis for seven to ten days. Buckeyes live two to three weeks as an adult butterfly, depending on temperatures and the environment.

**Plant type:** Perennial herb.
**Light:** Partial to full sun.
**Water:** Once established, low water needs and drought tolerant.
**Soil:** Needs good drainage but tolerates a wide variety of soils.
**Size:** 6–12 inches tall; spreads to 4 feet via runners.
**Bloom color:** White with red and yellow accents.
**Bloom time:** Spring through late fall.

**Availability: Widely available at nurseries and online. Fast grower. Also, easy to propagate by cutting off runners and rooting in potting soil.**

**Note: frogfruit is easy to grow. Don't scrimp on water when you're getting it established. Once it takes off, you can propagate it by snipping the runners with small roots and planting them elsewhere.**

Nobody knows for sure why it's called frogfruit, but some speculate the name evolved from "fogfruit," since the plant thrives in damp, moist soils along riverbeds and streams.[1]

According to ethnobotanical sources, raw frogfruit leaves can be eaten when young. The leaves can also be stewed as a tea, which has been described as having a grassy flavor. The root of frogfruit has been used as a treatment for hookworm, and research suggests the plant is loaded with phytochemicals containing antimicrobial, antibacterial, and antifungal compounds.[2]

In India, frogfruit is known as *poduthalai* and is famous as a cure for dandruff. (*Podugu* means "dandruff" in the Tamil language.) One recipe suggests boiling fresh frogfruit leaves in coconut oil until the water evaporates. "Switch off, cool and strain, and use it as a hair oil to get rid of dandruff."[3]

# FROSTWEED

## *Verbesina virginica*

Asteraceae / Aster family
Medicinal / Crazy ice crystals / Wildlife / Deer resistant

Frostweed pulls one of the most intriguing stunts in nature: crystallofolia.[1] Upon the first hard freeze, *Verbesina virginica* splits its sturdy, winged stems and pours ice crystals from its stalks. Airy, fragile constructions ooze from the plant like artisanal meringues. Touch these delicate structures and they break; breathe on them and they begin to melt. The first time I witnessed frostweed's ice sculptures, I couldn't help but think how urban mixologists could utilize the natural treasures as adornments on fancy adult beverages.

It's hard to understand why the plant has been so widely ignored by the commercial nursery business.

Frostweed is one of few plants to exhibit crystallofolia.

Frostweed provides nectar to butterflies in the fall.

In addition to its novel ability to produce ice sculptures, *Verbesina virginica* grows in sun or shade and is drought tolerant, low maintenance, and native to much of the United States. Gardeners in hot climates appreciate its ability to stay green in harsh summer heat. Its spectacular white flowers bloom from August through November. Its lush umbels serve as a prime nectar and pollen source for monarchs and other butterflies, as well as a favorite of native solitary bees and honeybees. Late-season nectar plants are prized by the wildlife community since many members are building up their fat stores for the winter.

While frostweed's mass of seeds provides a plentiful source of winter protein for birds, deer don't seem to care for it.

*Verbesina virginica* sports sturdy stalks with fleshy green flanges. They can reach four to eight feet in a year with good rain, or if watered regularly. The plant is also known as white crownbeard (presumably for its crown of white flowers), white wingstem, ice plant, and Indian tobacco—because Native Americans smoked frostweed as a tobacco substitute and used it in ceremonies.

Bees enjoying frostweed.

Monarch butterfly resting on frostweed.

Fortunately, the plant is available in seed form and germinates readily. Once established, it spreads through seeds and rhizomes and can be easily divided and shared or transplanted. It makes a great background plant, forming thick colonies under trees or in wildscapes. While frostweed is a perennial, it's also biennial—which means it will not flower until the second year.

**Plant type: Perennial or biennial herb.**
**Light: Partial sun to shade. Thrives under tree canopies with morning and early sun.**
**Water: Low to medium.**
**Soil: Well drained. Tolerates acidic to calcareous soils.**
**Size: 3–6 feet tall, 2–4 feet wide.**
**Bloom color: White.**
**Bloom time: August to November.**

**Availability: Seeds available through native plant seed suppliers; not widely available at nurseries. Once established in your landscape, easy to divide and transplant.**

The Choctaw pounded frostweed's roots and made a cold infusion to reduce fevers. The Chickasaw made tea from its leaves to address venereal diseases.[2] And the Seminole used it ceremonially and medicinally for everything from fever and chills to eye infections and stomach issues.

As for those ice sculptures, only a handful of species exhibit this little-studied trait. The first published scientific reference to the intriguing ice formations was made in 1824 about a different member of the aster family, one that is now called stinking camphorweed, *Pluchea foetida*, in Stephen Elliott's *A Sketch of the Botany of South Carolina and Georgia*.[3] More than one hundred years later, Robert T. Harmes, a linguistics professor at the University of Texas at Austin who dove into botany upon retirement, coined the term "crystallofolia" to describe the phenomenon.[4] The name comes from the Latin *crystallus*, "ice," and *folium*, "leaf."

If you happen to be near frostweed when the first frost arrives, go take a look early because the crystals melt quickly as the day heats up.

Ailanthus webworm moth on frostweed.

# GOLDENROD

*Solidago spp.*

Asteraceae / Aster family
Culinary / Cut or dried flowers / Medicinal / Wildlife

Sprays of vibrant yellow goldenrod usher in fall and make it a great late-season nectar plant during a time when most flowers are fading. Native plant advocate and biologist Doug Tallamy labels goldenrod "one of nature's greatest gifts to animal life." Its leaves support more than one hundred species of caterpillars in some parts of the United States, as well as numerous beetles. Its flowers provide pollen and nectar for thirty-five species of bees, fifteen of which are goldenrod specialists, meaning they will use *only* goldenrod pollen. Birds and mice eat its seeds.

    The plant is a magnet for bees and butterflies and a favorite of migrating monarchs as they move south in the fall, fueling up en route to their

Giant swallowtail on goldenrod.

Goldenrod can be incorporated into urban landscapes.

winter roosts. Goldenrod honey is prized for its bold taste and medicinal benefits. And its hollow stems make ideal homes for mason bees, carpenter bees, and other specialists. The straw-like stalks can be bundled and hung in the landscape to make a "bee condo" for solitary bees.

Goldenrod's fluffy yellow spikes add late-season interest to any garden and make beautiful cut flowers, as well. Harvesting flowers early in the bloom cycle helps them keep longer. Dried, the golden plumes can be woven into a seasonal wreath. Flowers can also be tapped to make a striking yellow dye for yarn or fabric.

*Solidago* boasts multiple medicinal and culinary uses, drought tolerance, deer resistance, ease of growth, and many benefits to wildlife.

Yet despite these many virtues, goldenrod has long been shunned by American gardeners as a "weed." Europeans have embraced it as a garden star for decades, but here in the United States, the plant has been historically misunderstood.

Goldenrod's PR problem results, in part, from a timing issue. The plant blooms at the same time as ragweed, one of the most common fall allergens.

In general, insects pollinate plants with flashy, attractive flowers like goldenrod, which produces a heavy, sticky pollen. Flowers with understated, nondescript blooms and lightweight pollen, like ragweed, let wind do the work—resulting in the itchy eyes, runny nose, and congestion associated with it.

But because of this blooming synchronicity, goldenrod takes the rap for ragweed's allergenic reputation. Ironically, native peoples used goldenrod medicinally to address seasonal ragweed allergies. Teas, tinctures, and infusions made from goldenrod's leaves and flowers provide decongestant relief, while ragweed remains one of the most common pollen allergies in the United States, affecting 15 percent of the population, according to the Asthma and Allergy Foundation of America.

Another trait that contributes to goldenrod's underwhelming reputation is how easy it is to grow. It can quickly colonize abandoned fields, roadsides, and flooded areas.

It spreads via seeds and rhizomes, which form after the first year and can be aggressive, depending on the species. *Solidago canadensis*, the version we have in Central Texas, can easily bloat from a selection of plants

Goldenrod is often inaccurately accused of causing pollen allergies.

one year, to a yard full the next. This can be a good or bad thing, depending on your goals. Personally, I'd rather have a living "mulch" such as rambunctious goldenrod than dead pine bark chips. It's also easy to pull out the young plants when you're ready to make room (and some tea?) for something else. You can also use a weed whacker to tame out-of-control goldenrod. In the meantime, your lush, low-maintenance landscape feeds wildlife, and goldenrod's roots fuel the soil.

If you have access to wild goldenrod, seedlings are easily dug up in the spring and transplanted. Enlightened commercial growers have caught on to goldenrod's charms and developed hybrids. Plugs of native and hybrid species are available through nurseries. Seeds are also an option, preferably sown in spring. Check your local native plant society or agricultural extension for help in deciding which *Solidago* species best serves your landscape.

**Plant type: Perennial herb.**
**Light: Sun to part shade.**
**Water: Medium.**
**Soil: Not picky. Grows almost anywhere—clay, loam, sand, caliche.**
**Size: 3–6 feet tall.**

Meet my living goldenrod mulch.

**Bloom color: Yellow.**
**Bloom time: Late summer to early fall.**
**Nuisance potential: Very rambunctious, but easy to pull out. Grows via seeds as well as underground rhizomes, which fuel the soil. Useful as a living mulch but can dominate.**
**Availability: Plugs and seeds available.**

The plant's Latin name, *Solidago*, means "to make whole" and suggests its well-rounded utility as a natural pharmaceutical. The more than one hundred species of goldenrod have a long history of medicinal use around the world.

Legend has it that goldenrod's name resulted when an old woman walking through the woods became tired, suffering from sore feet. She asked several trees for a walking stick, but they all refused to assist. The woman found a stick and asked it for help. The stick agreed to be her staff and she used it to make her way through the forest.

When the old lady emerged from the woods, she turned into a beautiful fairy. She asked the stick what it wanted most in the world. It replied that it wanted the love of children everywhere. The fairy responded by turning it into a flower, sprinkling it with gold dust and declaring it a "golden rod" to be forever loved by children.[1]

According to the Herb Society of America, Native Americans made goldenrod infusions to treat fever, kidney problems, and bladder ailments. The Chippewa called goldenrod "sun medicine," boiled it in water, reduced it to a syrup, and used it to treat respiratory infections. Indigenous peoples used various iterations of goldenrod to induce vomiting, for pain relief, and as "a charm for success in gambling," according to the Native American Ethnobotany Database. When colonists boycotted tea following the Boston Tea Party, a soothing "Liberty Tea" emerged, featuring goldenrod leaves as a chief ingredient. It was considered "a cure for melancholy."[2]

And in Brazil, a particular variety, *Solidago chilensis*, known there as "arnica," is widely used to soothe muscle aches and inflammation.

## GOLDENROD VINEGAR

Adapted from Edible Wild Food[3]

*This vinegar improves mineral balance and may help prevent kidney stones. Most of all, it adds a wonderful flavor to most salads.*

*Ingredients*

    2 cups chopped goldenrod leaves, flowers, and (cleaned) roots

    4 cups organic apple cider vinegar

    1 1-quart mason jar (sterilized)

*Directions*

    1. Place all goldenrod parts in the mason jar (be sure all parts are free of moisture). Pour vinegar into the jar and be sure to mix well to ensure no air bubbles are trapped.

    2. Place lid on, but be sure to have a piece of waxed or parchment paper to protect the vinegar from the lid. Not only can the vinegar erode metal lids, but older lids contain BPA.

    3. Label with date and contents.

    4. Let the mixture sit in a cool, dark location. Shake mixture several times a week. After 6–8 weeks, strain the contents and squeeze extra vinegar from the plant matter. Return vinegar to jar. Enjoy on salads and as a seasoning.

CAUTION!

*Arnica montana*, a different genus in the aster family, commonly known as arnica, is widely used in botanical medicine in the southern United States and is mildly toxic.

# HUMMINGBIRD MINT

### *Agastache spp.*

Lamiaceae / Mint family
Culinary / Medicinal / Great cut flowers / Wildlife

With its upright pink to purple flowers, easy-to-grow reputation, and tolerance for drought, *Agastache* offers an appealing package for the landscape.

This eye-catching perennial can reach two to four feet, depending on the species. Its blooming spikes host dozens of tiny flowers that cover the top third to half of the plant and create an attractive border, backdrop,

Anise hyssop, a type of *Agastache*, makes a dramatic statement and attracts bees. Photo courtesy of Baker Creek Heirloom Seeds, rareseeds.com

or group planting in the garden. With twenty-two species native to the United States, an *Agastache* likely exists to suit your situation.

Known as fragrant hyssop, anise hyssop, Mexican hyssop, giant hyssop, and hummingbird mint, *Agastache* and *Hyssopus* are actually two distinct genera in the mint family. As often happens, common names confuse their identities.

*Hyssopus* includes ten to twelve species and is native to the Mediterranean, while *Agastache* hails from North America. Both plants flaunt rows of tiny tubular flowers on their square-shaped stems. The leaves and flowers of both plants are especially fragrant and attractive to pollinators. *Agastache*, however, has earned a reputation in recent years as a hardy, drought-tolerant native that proves irresistible to bees, butterflies, and especially hummingbirds.

On a recent road trip out West, my husband and I chanced upon some beautiful *Agastache foeniculum* plants at a native plant nursery in Santa Fe. The species is the most common *Agastache* commercially available and is known by the common name hummingbird mint. Upon arriving at our Airbnb, we stored the one-gallon pots on the balcony before the drive home to Texas.

Within hours, hummingbirds abandoned the sugar syrup in the neighbors' commercial plastic hummingbird feeders to feast on the *Agastache* as we watched in delight through the glass door from inside our condo. The plants have done well in our urban pollinator habitat in downtown San Antonio, their flashy spikes luring a variety of pollinators.

Hummingbird mint, another type of Agastache. Photo courtesy of Baker Creek Heirloom Seeds, rareseeds.com

Its name, which is fun to say aloud and sounds like a congratulatory greeting, comes from the Greek words for "intense" or "extreme" and "ear of corn or grain"—presumably referring to the multiple rows of flowers on the plant's vertical stems.

The Xerces Society for Invertebrate Conservation views *Agastache* as especially beneficial to native bees and suggests planting the upright perennial next to soft prairie grasses to create contrast.

**Plant type: Perennial.**
**Light: Full sun / part shade.**
**Water: Extremely drought tolerant.**
**Soil: Not particular.**
**Size: 2–4 feet tall, spreads to 5 feet.**
**Bloom color: Orange, yellow, red, but wide variety, depending on species.**
**Bloom time: April to November, sometimes later depending on climate.**
**Availability: Available as seeds, as well as pots at nurseries. Can be divided in winter or early spring from the root.**

*Agastache* has a long history of medicinal use, and one of the first mentions of it is in the Chinese medicine bible, the *Divine Husbandman's Classic*, in 500 AD. The herb stimulates the digestive tract and has been

CAUTION!

Extreme caution is advised regarding the use of any wild herb during pregnancy where the chemical constitution is not completely known. It could be dangerous and is likely best avoided.

prescribed to address bloating, nausea, vomiting, and indigestion. It also has a history of treating morning sickness.

Numerous studies document *Agastache*'s unique and useful chemical makeup. One showed that an infusion made from Mexican *Agastache* leaves and flowers created an effective treatment for jaundice; another demonstrated that *Agastache* honey has superior antifungal activity compared to commercial honeys.[1] Not surprisingly, *Agastache* honey is sold as "wonder honey."

In a 1940 article in the *American Bee Journal* titled "Anise Hyssop, Wonder Honey Plant," the early twentieth-century beekeeping maestro Frank Pellett of Iowa detailed the many attributes of *Agastache*.

Pellett was trying to locate the plant, which he had read about in a 1920s dispatch authored by H. A. Terry, a plant pioneer and hybridizer of peonies. Terry had claimed that a single acre of *Agastache* could support one hundred bee colonies because of its long bloom cycle, which he cited as June through November. As a beekeeping expert, Pellett was intrigued.

"The old world hyssop which has been cultivated for centuries has been brought to American gardens but this one which was so much loved by the Indians has been permitted to disappear almost completely from its native region," noted Pellett in the article.

Pellett searched in vain for the plant to install it in the *American Bee Journal* honey plant test garden. After months of searching, someone north of Winnipeg, Canada, sourced a dozen plants.

Although the first seed trial was not successful, the second met with success. Pellett scattered a light bed of straw over the tiny seeds to keep them from drying out, and thousands of plants resulted.

The bees worked the *Agastache* "vigorously," wrote Pellett. A plant that blooms for months and can have more than one hundred flower clusters at one time, as *Agastache* does, "is certainly worthy of more attention than it has received,"[2] he concluded.

While *Agastache* is deer resistant, pollinators are not the only ones that feast on its charms. The foliage and flowers offer notes of mint, licorice, and anise when brewed as a tea and can be added to fruit or savory salads as an ingredient or a garnish. Pop a flower in your mouth for a minty, sweet taste. Fresh or dried, *Agastache* can be used to season meats or added to cookies, pies, or stewed fruit—anything that calls for a minty, licorice taste.

*Agastache* leaves and flowers are edible. Photo courtesy of Baker Creek Heirloom Seeds, rareseeds.com

*Agastache* also makes great cut flowers that hold up well and can be used in dried flower arrangements. Potpourri is another option, given its pleasant fragrance.

## LUSCIOUS AGASTACHE BROWNIES

This recipe is adapted from our friends at Central Market Cooking School in San Antonio. The Agastache adds a subtle floral, minty flavor to the rich, dark chocolate and nuts. These very moist brownies keep for several days and are actually more manageable and tastier after a rest.

*Ingredients*
   2 large squares dark chocolate
   ¼ cup butter
   2 eggs
   1 teaspoon vanilla
   ¾ cup flour
   ½ cup chopped hazelnuts, pecans, or walnuts
   2 tablespoons (or more to taste) chopped fresh *Agastache* leaves and flowers
   Powdered sugar for sprinkling after baking

*Directions*

1. Preheat oven to 325°. Rub a glass square baking dish with butter and dust with flour.

2. Melt chocolate and butter together and cool.

3. Whisk chocolate mixture into everything except the nuts and *Agastache*.

4. Stir until smooth, add nuts and most of the chopped *Agastache*. Reserve a bit to sprinkle on top of the brownie batter.

5. Pour into buttered, flour-dusted 8-inch square baking pan. Sprinkle reserved *Agastache* on top.

6. Bake for 25 minutes and dust with powdered sugar.

*Note: you can also jazz up a prepackaged brownie mix by adding Agastache to the batter. For an extra flourish, sprinkle whole flowers on top after baking.*

Chopped Agastache leaves ready for service.

Agastache-laced brownies.

# JIMSON WEED

### *Datura wrightii*

Solanaceae | Nightshade family
Medicinal | Night bloomer | Interesting history | Wildlife / Deer resistant

With its elegant white trumpet-shaped flowers, spiny seed capsules, and fragrant evening blooms, Jimson weed earns its slot in the garden. The native *Datura wrightii*, known as Jimson weed, Jamestown weed, angel's trumpet (for its flower shape), or moonflower (for its nighttime blooming habit), can take summer's brutal heat and is native to much of the United States, but common in the Southwest. Jimson weed was a favorite of the great American artist Georgia O'Keeffe, who memorialized it in an oil painting in 1936. The plant has many stories to share.

But first, Jimson weed boasts plant versatility and curb appeal, and it's easy to grow. *It's also highly toxic if consumed.* Requiring little water or

Jimson weed makes a fine addition to the pollinator garden.

care, it resists disease and pests and attracts bees and moths. Its dramatic white blooms emerge late in the day, flower fully at night, and give off an enchanting smell. Jimson weed climbs to three feet and spreads an equal distance. It creates a handy shady mass that protects less sturdy plants under its canopy.

What else does this member of the nightshade family have to offer?

Its spiny seedpods inspire one of its common names, thorn apple. The thorny green balls would make delightful earrings or could play a starring role in an exotic ikebana flower display.

Jimson weed seedpod.

Jimson weed seedpods eventually burst open and spill their seeds.

Jimson weed's white flowers bloom at night and attract moths.

As summer wears on, the walnut-sized pods turn from green to brown, burst open, and spread seeds wantonly in the garden, making this durable plant almost impossible to defeat once established. Described as both an annual and a perennial, depending on the species and growing conditions, Jimson weed forms a large tuber that seems to guarantee its survival. It can die back in a freeze and emerge hardy in the spring.

Its fuzzy, lush leaves are laden with glandular trichomes, specialized hairs that exude myriad chemicals and essential oils that account for Jimson weed's distinctive smell. Some folks compare it to the smell of a wet dog, but my nose perceives Jimson weed as chocolaty and distinctive. Perhaps it depends on the soil where the plant grows, or differences in perception. The chemicals and scents produced by the glandular trichomes have been associated with deterring insect predation,[1] which may explain why Jimson weed is so easy to cultivate.

While all parts of this plant are poisonous, the Carolina sphinx moth caterpillar feasts voraciously on Jimson weed's foliage to no ill effect. The large moth with pink stripes enjoys the fuzzy, smelly leaves as a caterpillar, and the fragrant, sweet nectar of the plant's trumpet-like flowers as an adult.

The Carolina sphinx moth caterpillar closely resembles that of the much-loathed tomato hornworm. The tomato hornworm, *Manduca quinquemaculata*, has eight V-shaped marks on each side of its green body and a signature horn on the rear. The caterpillar of the Carolina sphinx moth, *Manduca sexta*, shows seven diagonal white lines on its sides and a curved horn.

Both caterpillars adopt a sphinxlike pose when bothered or threatened, thus their name. And each morphs into a large moth with a four-to-six-inch wingspan in colors ranging from brown and gold to pink and gray. Since the moths are mistaken for small hummingbirds when they fly during the day and hover helicopter style to nectar on flowers, they are also called hummingbird or hawk moths.

CAUTION!

Underappreciated Jimson weed does have a dark side. As a member of the nightshade family, it contains the same tropane alkaloids found in belladonna and used in ancient times on poison-tipped arrows. *All parts of the plant are poisonous.* Deer and dogs avoid it. Native Americans used the leaves as a painkiller and hallucinogen. Adventurous teens have experimented with Jimson weed to get a cheap high—but they should beware. Hospital stays, even death, can result.

Early-stage sphinx moth caterpillar and egg on Jimson weed.

Late-stage sphinx moth caterpillar on Jimson weed.

**Plant type:** Annual herb.
**Light:** Sun to part shade.
**Water:** Low to medium.
**Soil:** Not picky; sandy, loamy, clayish.
**Size:** 3 feet tall, spreads 3–4 feet.
**Bloom color:** White.
**Fruit:** Thorny green capsule turns brown and spills reddish-brown seeds. Capsule and seeds are poisonous. Do not eat.
**Bloom time:** Spring through fall.
**Availability:** Seeds available and occasionally found in specialty nurseries.

Jimson weed's namesake may represent one of the first instances of ethnobotanical warfare in American colonial history. As Amy Stewart explains in her delightful book *Wicked Plants*, in 1607 in Jamestown, Virginia, settlers encountered a "seductive, beautiful weed."[2] Unfortunately, they consumed it and many perished, likely after horrifying bouts of delusions and convulsions.

About seventy years later, when British soldiers arrived to suppress rebellion in the fledgling colony, the settlers recalled the toxic plant and slipped Jimson weed leaves into the soldiers' food. The soldiers survived but suffered hallucinations for eleven days. That extra time gave the Virginia colonists a temporary upper hand. The assisting plant became known as Jamestown weed, and later, Jimson weed.

Centuries before that, Jimson weed occupied the spiritual and medicinal arsenal of native peoples in Mexico, where it's known as *flor de toloache*. Indigenous midwives used it to relieve labor pains of childbirth.[3] Even today, it has a reputation for working as both a love potion and a death sentence.

A friend from Mexico shared that when one of the partners in a couple is mean, dominating, and inconsiderate, and the other is noble, it's hard to understand why the good one stays in the relationship. The explanation? "We say it's because 'lo tiene entoloachado.'" In other words, the noble partner is "under a spell."

Cave paintings at Panther Cave along the Devils River in South Texas include Jimson weed "power bundles."

Near the confluence of the Pecos River, Rio Grande, and Devils River in South Texas, archaeologists at the Shumla Archaeological Research and Education Center in Comstock have suggested that ancient paintings that feature a thorny circular motif may represent Jimson weed seedpods. They call the motif a "power bundle."[4]

Scientists speculate that early peoples tapped the hallucinogenic properties of Jimson weed to connect with the gods. I like to incorporate a Jimson weed seedpod when wrapping birthday gifts, to provide friends and family with that extra spiritual jolt that comes with the passing of another year.

Jimson weed "power bundles" make a sustainable and novel gift decoration

Bees on Jimson weed flower.

## ONE FINAL WORD OF CAUTION!

All parts of this plant are poisonous, and consumption of it can be fatal. Jimson weed is highly toxic. It should not be used internally by the inexperienced. The adventurous have died from smoking or ingesting it.

# LANTANA

*Lantana urticoides, Texas lantana*

Verbenaceae / Verbena family
Drought tolerant / Medicinal / Deer resistant / Wildlife

Loves poor, dry soil. Needs little water. Thrives in extreme heat. Attracts butterflies, bees, and hummingbirds to its nectar and pollen—but deer won't touch it. It's the host plant to the lantana scrub-hairstreak, birds eat its dark purple berries, and its leaves provide medicinal benefits. What's not to like about lantana?

Lantanas come in many colors. Try to buy native varieties.

Well, it can be a bit rambunctious if you plant the wrong species. But choose wisely from the more than 150 species available, and you'll find lantana a congenial resident in the landscape.

Besides heat tolerance and a carefree manner, this low-growing shrub, also known as Texas lantana, calico bush, and shrub verbena, is noted for its prolific flower power. *Lantana urticoides* blooms nonstop from April through fall and produces dozens of red, orange, and yellow one-to-two-inch groups of flowers. A nonnative variety from the tropics, *Lantana camara* (not a good candidate for most gardens even though it's widely available), flaunts pink and yellow flowers and inspired the cultivar much promoted by commercial nurseries: 'Ham and Eggs' lantana.[1]

Lantana sports rough, scratchy leaves, and while some varieties can smell like mint, many people (not I) find its fragrance unappealing, thus its original botanical name, *Lantana horrida*, or "horrid." Austrian botanist August von Hayek renamed it *Lantana urticoides* sometime in the late nineteenth century, yet the "horrid" name still lives on in books and online.

The multicolored flowers, one of lantana's most appealing traits, later form green berries that eventually turn dark purple or black. Birds love the ubiquitous fruits and are immune to the triterpenoids, one of many potentially toxic and distasteful chemicals contained in lantana berries.

While lantana is native to the Americas, and *Lantana urticoides* is found naturally from Texas to Arizona and into southern Mexico, the plant has been widely cultivated and hybridized. It was imported to Europe in the 1800s and captivated gardeners there with its abundant flowers and dark green foliage. As often happens, gardeners selectively bred particular plants with particular features, and today myriad cultivars and crossbreeds exist.

Some lantana varieties have become invasive in areas where circumstances allow them to dominate the ecosystem. *Lantana camara*, native to the West Indies, Colombia, and Venezuela, has been identified as an especially onerous invader, displacing native plants with its rambunctious growth and domination of resources, especially in disturbed areas. Birds have facilitated its spread by dropping the digested seeds consumed via the berries along their flight paths. The problem became so great that plant breeders developed sterile varieties with names like 'Hot Blooded,' 'New Gold,' 'Alba,' 'Patriot,' and the 'Bloomify' Series.[2]

## CAUTION!

Consumption of lantana fruit can be lethal. Some reports say that only green fruit is harmful, but children, sheep, and calves have died from eating the leaves and fruit. The rough foliage can also cause a skin rash in sensitive individuals.[3]

Young lantana drupes, or fruit. Do not eat them.

That said, our dogs Cocoa, Cacteye, and Brisket have all gravitated to lantana for stomach purges. They eat a few leaves, then upchuck whatever was bothering them. As stated in other chapters, bitter flavors and strong smells typically accompany toxic plants (milkweed, red yucca, Jimson weed). While a dog might be uncomfortable for a bit after chewing a few lantana leaves, the damage has never been serious in my experience. Keep in mind that the species, the type of soil in which the plant is grown, and the climate conditions in addition to a particular dog's or creature's sensitivities and body chemistry can affect what impact a taste of lantana or other potentially toxic plant might have. If you have concerns, keep your dogs, children, and other mammals away from lantana.

In my hometown of San Antonio, local authorities specifically discourage the planting of *Lantana camara* for its invasive tendencies, especially along the San Antonio River. Our municipally owned water utility, a leader in water-saving gardening, has also called out another nonnative species, *Lantana montevidensis*, for its overaggressive growth habit and domineering style. Known as trailing purple lantana, this species was cultivated to be sterile.

Yet, as stated on the San Antonio Water System's Garden Style San Antonio website, "This South American lantana was once thought to be seedless. But in the spirit of 'nature finds a way,' it has been found spreading into natural areas in Central Texas, so it is no longer recommended."

That's why it's important to choose a native lantana or one that is well adapted to your area. The plant has much to recommend it with its ease of growth, low maintenance, and wildlife-friendly appeal.

Bees, butterflies, and hummingbirds gather on lantana. Its short-tubed flowers provide easy access to nectar, and its continuous bloom cycle makes its sugary fuel available much of the year.

The rare lantana scrub-hairstreak butterfly, *Strymon bazochii*, found in the Rio Grande Valley of Texas and many points south, utilizes Texas lantana as fodder in its caterpillar stage. The common gray hairstreak butterfly, *Strymon melinus*, is reported to use lantana leaves as a larval food source, as well.[4] I have never personally witnessed this, perhaps because the common hairstreaks are not particular about their food as caterpillars and consume a variety of plants in their larval stage. I have witnessed gray hairstreak butterflies nectaring on lantana many times, however.

Drought-tolerant lantana is a favorite landscaping plant in hot climates.

Deer avoid lantana, most likely because of its strong smell. Its scratchy leaves also deter grazing, which is why for landscapes with deer populations, lantana is a good bet.

Lantana can reach two to three feet in height and two to six feet in breadth. It loses its leaves in winter but recovers its full vigor in spring. For the best flower show, cut it back in winter, as the flowers form only on new growth.

Plant type: Low-growing perennial shrub.
Light: Full sun / part shade.
Water: Extremely drought tolerant.
Soil: Not particular.
Size: 2–4 feet tall, spreads to 5 feet.
Bloom color: Orange, yellow, red, but wide variety, depending on species.
Fruit: Green berries that follow flowers turn dark purple or black. Toxic. Do not eat.
Bloom time: April to November, sometimes later depending on climate.
Nuisance potential: If you choose the wrong species, it can dominate. Some lantanas have been officially declared "invasive" in certain regions. Do your homework to avoid problems.
Caution: Don't eat leaves, flowers, or berries.
Availability: Widely available at nurseries. Can be divided in winter from the root.

Various species of lantana extracts have been used in folk medicine for centuries to reduce fever and to treat cancer, chicken pox, measles, asthma, ulcers, swellings, eczema, tumors, and high blood pressure.[5] An ethnographic paper written in the 1930s on the medicinal uses of lantana cited the Warihio Indians of the Sonora-Chihuahua using a decoction of lantana to treat insect stings and snakebites.[6] And a 2014 study explored the potential of developing a new biopesticide from the leaves of *Lantana camara*.[7]

# MARIGOLD

*Tagetes spp.*

Asteraceae / Aster family
Culinary / Great cut flowers / Insect repellent / Interesting history /
Medicinal / Wildlife

The worldly marigold has a convoluted history. Native to Mexico and Guatemala, *Tagetes* species have been used as food, cures, and a connection to the gods since ancient times. When the conquistadors arrived in Mexico from Spain in the 1500s, they became enamored with its bright yellow and orange pompom blooms. Native marigold seeds were taken from the Aztecs to Spain. They ended up in monastery gardens.[1]

Soon, "French marigolds" and "African marigolds" were taking root. But they all derived from the original Aztec species, *Tagetes erecta*, one of about fifty. European gardeners found its showy flowers so enchanting, they cultivated it heavily. Marigolds made their way back to the Americas after the Revolutionary War as one of many plants shipped from Europe to the United States.

Today, low-maintenance marigolds with their vibrant, long-lasting blossoms have made *Tagetes* one of the most popular plants in the world. In 2018, Fleuroselect, an international trade association of the ornamental plants industry, declared it the plant of the year. The marigold is used in its native Mexico each November during Day of the Dead (*Día de los Muertos*) celebrations. The bright flowers and strong scent are said to guide the dead to their respective altars. And in India, weddings, festivals, and religious ceremonies often incorporate marigolds, even though the plant is believed to have arrived there as recently as the seventeenth century.[2]

Why is this plant so widely embraced?

Monarch butterfly resting on a marigold.

First, marigolds are famously easy to grow from seed or transplants. With full sun, decent soil, and water when dry, they pretty much take off. Their sunny blossoms have a long bloom season, often extending for months. Varieties exist to fit every situation—short, tall, compact, gangly, late bloomers, early risers. White, maroon, even striped marigolds have been developed.

Marigolds also attract bees and butterflies. Native species that have not been overhybridized make the best buffet for pollinators. My favorite is *Tagetes lucida*, also known as Mexican tarragon or Mexican mint marigold. While its golden flowers are more petite than those of other species, its fragrant serrated leaves can be used as a seasoning or tea. And those autumn blooms provide late-season fuel for pollinators.

Not all insects are attracted to marigolds. Thanks to limonene, a chemical contained in the plants' biomass, they exude a smelly odor. The smell, coupled with marigolds' easy growth habit, caused some early twentieth-century gardeners to dismiss them as a "stinky weed."

That "stink" can be a good thing. Limonene repels mosquitoes and other insects, like whitefly, a major pest in the commercial tomato industry.

Gardeners are often advised to plant marigolds near seating areas to repel mosquitoes or place them close to tomatoes as a companion plant to confuse the small, mothlike whitefly. The whitefly feeds on plant sap, transmits plant viruses, and encourages mold growth. Some have suggested that marigolds' ability to ward off insects makes it ideal for ceremonies involving the dead.

In a 2019 study, a team of researchers posed the possibility of using marigolds to develop a type of air freshener that could be hung in greenhouses to confuse whiteflies by exposing them to a blast of limonene.[3]

**Plant type: Annual herb.**
**Light: Full sun. If grown in shade may develop powdery mildew.**
**Water: Medium. Allow to dry out between waterings.**
**Soil: Not picky. Grows in most soils.**
**Size: 1–3 feet tall, depending on species.**
**Bloom color: Yellow and orange, but commercial growers have developed varieties in many colors.**
**Bloom time: Summer to fall.**
**Availability: Seeds, plugs, and containerized plants widely available.**

The first ethnobotanical reference to marigolds can be found in the *Códice de la Cruz-Badiano*, 1552, which is considered the first illustrated survey of Mexican nature produced in the New World. It prescribed *Tagetes* as a treatment for being struck by lightning. The survey also cited marigolds as a cure for hiccups, heat exhaustion, rectal swelling, scorpions, and more.[4] A sweet, licorice-flavored marigold tea is brewed to this day to address fever, stomachache, muscle pain, and other maladies.

In addition to their long-lasting beauty and medicinal, wildlife, and practical uses, marigolds have culinary value. Flower petals can be used as a natural dye to jazz up cookie dough, rice, teas, or soups. They add a subtle licorice flavor. Marigold petals are sometimes added to commercial chicken feed to make egg yolks more golden and appetizing; they're also mixed in with butter and cheddar cheese to boost color. Whole flowers can be deep fried like squash blossoms or tossed in a salad.

Marigolds hold special meaning for people all over the world.

In areas with large Mexican American populations, marigolds are used in the fall to celebrate the Day of the Dead (*Día de los Muertos*). The limomene "stink" referenced earlier had an entirely different meaning to the ancient Aztecs, who called marigolds *cempasúchil*. Indigenous peoples believed that the bright orange color of the flower and the distinctive scent led those who had died back to earth for a visit; thus *cempasúchil* became a preferred decoration at cemeteries and Day of the Dead celebrations.

# MEXICAN HAT

## *Ratibida columnifera*

Asteraceae / Aster family
Great cut flowers / Medicinal / Wildlife / Deer resistant

Mexican hat cuts a distinctive profile in the landscape. Its droopy rust or yellow petals surround a petite, conical seed head and create the silhouette of a broad-brimmed sombrero. The look of the seed cone inspires other common names for *Ratibida columnifera*—upright prairie coneflower, long-headed coneflower, and thimbleflower.

Caterpillar on Mexican hat in the Texas Hill Country.

Mexican hat seeds are easy to harvest.

Mexican hat seeds dry out as the season wears on.

In addition to its novel appearance, this drought-tolerant native is especially appealing to bees, beetles, birds, and butterflies. The namesake seed crown, again, is the draw.

It results from hundreds of tiny flowers that burst from the cone, a signature trait of the aster family and its composite flowers. The myriad tiny blooms can easily be mistaken for thick pollen granules, but each is a complete flower on its own. Pollinators generally favor composite flowers—and why wouldn't they? Bees and butterflies can sit in one spot and slurp nectar from multiple flower heads without having to budge.

After the flowers go to seed, birds feast on the protein-rich morsels. The flowers bloom from June through the fall, and the seed heads can stand for weeks, even months thereafter. Thus Mexican hat provides food for wildlife for much of the year.

Several moth species also feed on the seven *Ratibida* species in their caterpillar stages, consuming mostly the rays and florets. Some moth larvae feed on the roots.

Perhaps the most striking is the blackberry looper moth, a greenish moth with golden-tan markings that span its wings horizontally. A light gold fringe outlines the entire wing, calling to mind an understated Oriental rug. The blackberry looper is not a specialist, meaning it consumes other plants in its caterpillar stage, mostly those in the aster family. Its unusual coloring makes it an interesting visitor to the nonagricultural landscape.

Seven different *Ratibida* species display various color combinations and cone sizes, with rays of equal diversity. Mexican hat shows a more elongated cone surrounded by rust or yellow rays. *Ratibida pinnata*, known as gray-headed prairie coneflower, flaunts longer, yellow rays around shorter, rounder cones.

On top of its attractive appearance and benefits to wildlife, Mexican hat and other *Ratibida* species are easy to grow as well as deer resistant. They prefer full sun and well-drained soil and can be sown from seeds or planted from plugs.

If grown from seed, know that it will not flower until year two. After that, you'll pay little heed to this tough garden tenant, which reaches one to three feet in height with little water or fertilizer. The plant will reseed

Mexican hat can form impressive stands when conditions are right.

and return annually. It spreads freely and assimilates well. Use Mexican hat as a border, focus area, or member of a pocket prairie.

The seeds of Mexican hat are easy to harvest. After the cones have dried, simply snap off the head and roll it between your thumb and fingers. Dozens of clean Mexican hat seeds will fall into the palm of your hand. No chaff or fluff to clean or separate from the seed. Store in a cool, dry place for the next season or share with friends. The seeds also have a refreshing licorice scent that would be appropriate for a potpourri in the linen drawer.

As a cut flower, Mexican hat makes a strong impression as a group. A half-dozen Mexican hat blooms cut and poised in a vase make for an eye-catching bouquet. *Ratibida* species have a reputation for long-lasting cut flowers.

**Plant type:** Perennial herb.
**Light:** Full sun.
**Water:** Low to medium.
**Soil:** Well drained, not picky—from calcareous to clay to loam to acidic.
**Size:** 1–3 feet tall, about 1 foot wide per plant. Can form colonies.
**Bloom color:** Orange, yellow, brown.
**Bloom time:** May to October.
**Availability:** Seeds widely available and easy to propagate.

Native peoples used Mexican hat and other *Ratibida* species as teas, decoctions, infusions, and poultices. The Northern Cheyenne called Mexican hat "rattlesnake medicine" and boiled its leaves and stems, creating a yellow solution to apply to rattlesnake bites that would draw the venom from the wound.[1] A similar solution was made to treat poison ivy. Teas made from flower tops and leaves were consumed to relieve headaches, stomachaches, cough, fever—even epileptic fits. Boiled flowers were used to make an orange-yellow dye.

# MILKWEED

*Asclepias spp.*

Apocynaceae / Dogbane family
Medicinal / Monarch butterfly magnet / Wildlife / Technically edible

If you want to attract the most iconic insect in the Americas to your landscape, plant milkweed.

*Asclepias* species serve as the singular host plant to monarch butterflies, the storied international travelers that migrate from Mexico through the United States to Canada and back each year. Their multigeneration migration and role as an ambassador for all pollinators have fascinated scientists for centuries and made them one of the most studied and beloved insects on the planet.

Monarch caterpillar on swamp milkweed in a downtown San Antonio garden.

Note that milkweed is poisonous in varying degrees, again, depending on the species. Its family moniker, Apocynaceae, translates to "Away, dog!" or dogbane, because the plant family can be poisonous to dogs. The namesake milky latex that oozes from its stems and foliage contains cardiac glycosides and other toxins that can be irritating to the skin or burn the eyes. If consumed without proper preparation, milkweed can cause stomach and other problems—yet after it is boiled multiple times to rid the plant of its toxins, it can be eaten like asparagus. More on that to follow.

Though technically not in the *Asclepias* genus, milkweed vine, *Matelea reticulata*, is one of many underappreciated monarch host plants.

Those same toxic chemicals protect monarch butterflies and other milkweed-feeding insects from predators. The caterpillars ingest the toxins and sequester them in their bodies. This results in the butterflies' bright orange-and-black "warning" colors and advertises to predators, "Beware! Don't eat me!"

Monarch scientist Lincoln Brower popularized the understanding of this trait, known as aposematism, with his famous "Barfing Blue Jay" experiment. Brower fed monarch butterflies to a blue jay and soon the bird began to wretch, coughing up the bitter butterfly and presumably learning a lesson to never eat another monarch again.

If you have milkweed, be prepared to have aphids—here, on tropical milkweed.

With complex flowers in a range of colors—white, orange, yellow, and pink, depending on the species or variety—milkweed works as a low-maintenance perennial.

Also worth mentioning: milkweeds attract three species of aphids.[1] Sometimes known as plant lice, these soft-bodied sapsuckers undermine the plant but don't kill it. They produce honeydew, which attracts wasps, ants, and other insects, but aphids can be managed with strong sprays of water or a deliberate squish between thumb and fingers.

Milkweeds' multiple flowers serve as an outstanding food source for hover flies, hummingbirds, beetles, and bees. From seeds to stems, eleven different insects utilize milkweed exclusively, creating what chemical ecologist Anurag Agrawal has described as "the milkweed village."

Milkweeds also serve birds. Their elongated seedpods start out green in late summer and generally turn brown in the fall. When mature, the pods burst open to reveal flat, rust-colored seeds surrounded by fluffy white silk known as floss. The buoyant, hollow strands catch the wind and carry the seeds to new locations. They also make excellent nesting material for birds.

But milkweeds' biggest draw, besides their stunning flowers, remains the monarch butterfly. Females lay their tiny pearl-colored eggs on the plants in the spring. Shortly thereafter, a tiny whitish caterpillar emerges, consuming the eggshell as its first meal. As the caterpillar begins to ingest the chemicals from the milkweed leaves, it takes on the colors of a green, black, and yellow-striped suit, morphing through its stages before turning into a jade-green chrysalis with gold specks. The transformation from egg to butterfly takes about a month, and witnessing it is reason enough to plant milkweed.

More than one hundred species of milkweed exist in the United States, but most are commercially unavailable as plants. Depending on the species appropriate for your location, native milkweed seeds can be persnickety to grow.

Some species require stratification, or the wetting and chilling of the seed hull for specific durations to break it from dormancy; others prefer extremely calciferous or acidic soils. Most native species take two years to reach a marketable size and stature to sell commercially, which explains why growers have not cultivated them widely. Availability has improved in recent years, however. Pop-up plant sales staged by native plant societies and gardening groups are often the best bets to find locally appropriate milkweeds.

Milkweed's heavy chemical constitution makes it deer resistant as well. It can be gregarious, spreading by wind and via rhizomes. Ranchers don't like it because it can be poisonous to livestock, and farmers sometimes resent its aggressive growth habit, as it can crowd out cash crops.

Antelope horn milkweed is among the early risers in spring, providing host plant fodder for monarchs and nectar for other insects.

Tropical milkweed, *Asclepias curassavica*, has a reputation for its beautiful orange or yellow blooms, its ease of growth, and its commercial availability. Technically, tropical milkweed is not native to the United States. Its provenance lies south of the border, likely in Mexico

Ladybugs also gravitate to milkweeds.

Monarch butterfly on swamp milkweed.

Monarch chrysalis on swamp milkweed.

or Curaçao, and it's a favorite of monarch butterflies. In warm climates, the plant can thrive throughout the winter, and scientists believe its appeal can cause monarchs to break their migratory cycle. Its resilience in warm climates, attractive blooms, and available foliage can also result in a buildup of OE (*Ophryocystis elektroscirrha*), a common spore-driven disease that can be deadly to monarch butterflies.[2]

Thus, if you live in a warm climate where tropical milkweed thrives year-round, follow best practices as per monarch conservation organizations and cut the plant down each fall to about six inches. It will regrow in the spring.

> **Plant type:** Perennial herb.
> **Light:** Full to partial sun.
> **Water:** Highly variable, depending on species. Swamp milkweed, *Asclepias incarnata*, grows along rivers and requires more water. Antelope horns, *Asclepias asperula*, grows in rocky dry fields and needs much less.
> **Soil:** Good drainage, but again, depends on species.
> **Size:** 2–5 feet.
> **Bloom color:** Pink, white, green, orange, depending on species.
> **Bloom time:** Late spring through summer, depending on species.
> **Nuisance potential:** Tropical milkweed, *Asclepias curassavica*, is considered invasive and undesirable in some areas.
> **Caution:** Milkweeds contain cardiac glycosides and can be toxic to pets, people, and other creatures if consumed. The milky latex can irritate skin and eyes, so handle with care.
> **Size:** 1–3 feet tall, 1 foot wide.
> **Availability:** Butterflyweed *(Asclepias tuberosa)* and tropical milkweed *(Asclepias curassavica)* are generally available in nurseries via plugs; many other milkweeds are available via seed. If you want to stick to natives, I've found swamp milkweed *(Asclepias incarnata)* the easiest to grow from seed.

Note: if you grow tropical milkweed in a southern climate, please observe best practices and cut it down in the fall so it does not become a breeding ground for the deadly monarch-centric disease *Ophryocystis elektroscirrha*, commonly known as OE. This spore-driven disease kills

and causes deformities in monarchs and other milkweed-feeding butterflies, especially late in the season as the spore volume increases. This
could negatively impact the migratory population.

In his popular 1962 foraging book, *Stalking the Wild Asparagus*, Euell
Gibbons waxed poetic about common milkweed. He considered *Asclepias
syriaca* a multipurpose, overlooked feral vegetable. "Its wide distribution,
abundance and ease of procurement could make the milkweed an important wild vegetable if more people knew the secrets of processing its
products into palatable food," wrote Gibbons, devoting a chapter to the
plant. That processing includes multiple boilings of plant parts, followed
by dumping and replacing the water.

He described "wild parties" at which he served a crowd favorite—
unopened common milkweed buds, which he prepared by boiling the
young stems three times, each time tossing the milkweed liquid, and then
seasoning with salt and butter.

Native peoples used milkweed for various medicinal and practical
purposes. The Cherokee and Chippewa used *Asclepias* root decoctions
as a contraceptive as well as a treatment for backaches, warts, and heart
failure. In Africa, hunters tapped the poison of milkweed extracts for use
on the tips of their arrows to kill game and in warfare.

During World War II, milkweed floss was cultivated for use as filler
for life jackets. It's also been harvested as a supplemental stuffing for pillows, comforters, and jackets.[3] One company, Ogallala Comfort, has even
developed a line of luxury bedding products—pillows and comforters
stuffed with milkweed fluff.

Milkweed floss serves birds as nest
material.

# OAK

*Quercus spp.*

Fagaceae / Beech family
Wildlife / Shade / Culinary / Landscape mulch

If you can plant only one tree, let it be an oak. This mighty hardwood, celebrated in poetry, mythology, and history, is one of the most adaptable and long-lived trees on earth. The famous Major Oak of Sherwood Forest, where Robin Hood's merry band camped and cowered, is believed to be 800–1,100 years old. Over multiple centuries, it has

The renowned Alamo Oak in San Antonio, Texas.

127

witnessed Vikings, monarchies, Dickens, and Darwin and provided food and shelter to innumerable insects, birds, bats, and mammals as an overachieving contributor to the ecosystem.

Such is the nature of oaks. And with more than 550 *Quercus* species globally, an appropriate oak likely exists that will fit your landscape needs.

Why oaks?

Oaks are considered keystone species—that is, foundational to a healthy ecosystem. Just like the center rock in a Roman arch, a keystone species is essential to stability and health. Without it, the structure crashes.

In his book *The Nature of Oaks*, Doug Tallamy, a biologist and native plant advocate, makes the case that oaks, more than any other living organism, excel at transferring energy from the sun to plants, animals, and humans. They sequester carbon in their roots, reduce water runoff by soaking up rain and keeping soils moist, and feed and shelter wildlife and humans. Tallamy labels oaks as the most productive plant in the United States, with no other genus coming even close.

The polyphemus moth uses the oak tree as a host plant.

Oak foliage works as food for the larval stages of more than nine hundred species of moths and butterflies. You think caterpillars are gross? Recast them as bird food, since they serve as primary nutrition for dozens of bird species—birds whose babies rely on high-protein, easy-to-digest larval snacks to grow and thrive. It takes six thousand to nine thousand caterpillars to feed a single clutch of chickadees, and no plant provides more opportunity for larval production than an oak tree. Those larvae that make it to the moth or butterfly stage grace our gardens, pollinate our plants, and serve as food for bats, birds, and other creatures. All thanks to oaks.

The stunning polyphemus moth is just one example. This gigantic brown-and-black silk moth emerges in spring and summer, depending on where you live. It sports a wingspan of four to six and a half inches (one of the largest in North America), and intimidating owl-like eyespots rimmed with gold. In the caterpillar stage, it consumes oak leaves as well as those of sycamore, birch, and other hardwoods and grows to a chubby three to four inches long before forming a silk-bound cocoon wrapped up in a leaf right in the oak tree. The cocoon can remain in the tree, or if it's late in the season, it may drop to the ground, where it will wait for spring before it emerges as a moth. Polyphemus moths serve as primary fodder for bats, birds, and squirrels from Canada to Mexico.

But aren't oaks slow growers? Not really. Even in an oak's early years, when growth seems slight and unhurried, an acorn's few true leaves do not betray the massive machinations occurring under the soil. An oak sapling may appear only a few inches tall during its first year, but below the ground, a seedling oak may have up to ten times more root mass than the leaves and shoots above the earth. These root systems continue to grow over the oak's lifetime, providing soil stabilization, carbon sequestration, and water retention. As the tree flourishes, it blocks wind, shades your house, and effectively reduces heat island effects—and probably, your air conditioning bill. Planting trees and creating shade saves electricity. If you want to combat a warming planet over the long haul, an oak tree is one of the best investments you can make.

And what about all those persistent leaves that oaks drop in the fall (and sometimes in spring and summer if stressed)? People complain that oak trees are messy and their tannin-filled leaves take forever to decompose. That can be a good thing.

Native, organic, long lasting, and free: oak leaf mulch.

In places with alkaline soils, the acidic oak mulch fuels the soil. Given that oak leaves can take several years to decompose, your oak tree's fall leaf harvest provides you with ecozone-appropriate mulch at no cost. Rather than bag those leaves up and have them carted to the landfill, then have to spend money on commercial mulch, allow those leaves to over-winter on the ground where they will serve as a soil blanket and builder, as well as shelter for overwintering insects.

Beyond these benefits, acorns are a nutritious and tasty food when gathered and prepared properly. While acorn production varies depending on conditions, climate, and oak species, when the acorns drop, they can be gathered and stored for months, even years, when handled properly.

**Plant type:** Tree.
**Light:** Full sun.
**Water:** Low to medium.
**Soil:** Well drained, not picky—from calcareous to clay to loam to acidic; depends on species.
**Size:** 15–100 feet, depending on species.
**Bloom color:** Dangling tassels of tiny yellow-green flowers called catkins.
**Bloom time:** Spring.
**Availability:** Plant an acorn or buy a small tree at your local nursery.

Before corn and wheat became our go-to grains, indigenous people around the world gathered, stored, and utilized acorns as food. Sometimes, you can even eat acorns fresh, off the ground. Henry David Thoreau found them "unexpectedly sweet and palatable" and compared their taste to that of chestnuts, noting, "No wonder the first men ate acorns."[1]

Thoreau's tasty acorn was likely from a white oak. Oaks are segregated into "white" oaks and "red" oaks. White oaks generally produce acorns with lower tannin content that are typically more palatable as a fresh option. One way to tell white oaks from red oaks is leaf shape: white oaks generally have rounded leaf tips, while red oaks have pointed leaf tips. White oaks generally grow much more slowly than reds, as well.

In her book *Eating Acorns: Field Guide—Cookbook—Inspiration*, author and "oakmeal" activist Marcie Lee Mayer waxes nostalgic and poetic about the lost appreciation of acorns as food for people.[2] Citing their high iron, protein, healthy fat, and antioxidant content, Mayer calls acorns "a vital nutrition source for thousands of years" and includes recipes for acorn grits, muffins, and soup as well as acorn tortillas, brownies, and cookies.

Just like the trees that shed them, acorns come in various shapes and sizes.

Acorns are making a comeback in foraging circles. A recent *Wall Street Journal* article detailed acorns' newfound status as a "superfood" that can fight diabetes and obesity. In Seoul, Korea, the popularity of acorn noodles, jelly, and powder has exploded in recent years, challenging the squirrel population there as people forage all the acorns, leaving the squirrels starving.[3] The problem inspired the formation of the Acorn Rangers, a group of wildlife defenders that patrol wooded areas and call out illegal foragers. The Rangers also hide stashes of acorns under leaves for the squirrels to find.

Accumulating all that nutrition does require effort. Foragers must collect the oak nuts and then crack, peel, and (depending on the species) leach them to make a nutritious acorn flour. Some species of oak contain higher levels of bitter tannins, which require soaking the shelled acorn meat in a bath of running water for multiple days to remove the unpleasant taste to make the acorn meal taste good.

The lower the tannin content, the sweeter the acorn. The process can take days, weeks, or months, depending on the species. For the best-tasting acorns, arborist friends recommend the bur oak, *Quercus macrocarpa*, which has a low tannin content and enormous acorns with conspicuously fringed caps.

Bur oak acorns are enormous and bountiful. Photo courtesy of Aimee Holland.

The fruit of the oak has a long worldwide history of culinary use, dating back to Paleolithic times.[4] In the Americas, almost every indigenous tribe utilized the ubiquitous nuts as food, collecting and storing acorns for use during winter as acorn mush, gruel, oil, and flour for making bread. The Cahuilla tribe of California considered acorn meat a delicacy and used it at social and ceremonial occasions. Indigenous children even used acorns as toys, juggling with them or using them as jacks in games. Acorns were also strung as necklaces.

The nuts weren't the only oak outputs utilized. Oak leaves were used as bedding or were burned. Their ashes were tapped as a cleaning agent similar to lye. Pulverized oak bark was dusted on open sores as a healing agent and used to treat newborns with bleeding navels. Oak bark was also boiled and its infusions, concoctions, and teas were used to alleviate muscle pain, diarrhea, cholera, and bleeding piles. The boiled bark was also tapped as a dye for buckskins. Even the branches of oak trees were woven into work baskets as handles or used as a structural support.

And since Roman times, oak galls—weird tiny growths that look like small beads on the undersides of oak leaves, formed when oak gall wasps lay their eggs—were used to make ink. Tannic acid is a primary ingredient in ink. Oak gall ink was used to write the Declaration of Independence.[5]

# PASSIONFLOWER

*Passiflora spp.*

Malvaceae / Mallow family
Culinary / Medicinal / Showy fruits / Wildlife

Showy tendril-laced flowers, colorful, edible fruits, and a tendency to jump fences and property lines paint a rambunctious picture of passionflower. This gregarious vine comes across as an extreme workhorse in the landscape.

Birds love its fruits and it hosts several species of butterflies on its fast-growing foliage. Passionflower pollinators include bats, hummingbirds,

Passionflower's blooms, vines, and fruit always make an impression.

butterflies, wasps, and bees. Easy to grow, passionflower boasts dramatic beauty, a great family history, and edible fruits.

For example, *Passiflora lutea*, yellow passionflower, one of more than five hundred *Passiflora* species, serves as the only larval food source for a small, black ground-nesting bee known as the passionflower bee, *Anthemurgus passiflorae*.[1] The tiny solitary bee will only feed its larvae pollen from this wild yellow passionflower. Luckily for the passionflower bee, yellow passionflower is not uncommon and its range extends from Texas east to North Carolina and north to Illinois.

*Passiflora ciliata*, commonly known as fringed passionflower, grows profusely in Texas and Florida. Its native status is open to discussion, but native or well adapted, the plant has much to offer.

Its dramatic lavender and purple flowers, graced with white and maroon tendrils, emit a lovely fragrance and produce a puffy, bright red berry from late summer through November. In mild climates, the vines with red fruits make attractive holiday decorations, draped on tables or as gift garnishes. The fruit has little taste—it's mostly air. Crunchy seeds occupy a hollow magenta fruit the size of a Ping-Pong ball. The seeds are clothed in a gelatinous wrapper.

Also in the Malvaceae family: *Passiflora foetida*, sometimes known as corona de Cristo or stinking passionflower because of its smelly bloom.

You might think that the name "*corona de Cristo*"—crown of Christ—comes from the trinity of stigmas that leap from the flower's center. But *corona de Cristo* has its own narrative, cultivated by Spanish Christian missionaries in the 1600s. They interpreted the *Passiflora* flower as a metaphor for Christ's death and assigned each part of the complex flower a role in the crucifixion.

Passionflowers wear coats of many colors. Here, *corona de Cristo* wears white.

Purple passionflower.

The colors and fruit of passionflowers can vary greatly, but they all share dramatic flower structure.

According to multiple sources, the leaves represent the hands of Christ's persecutors, while the tendrils stand for the whips used to flagellate him. The five sepals and five petals symbolize the ten faithful apostles (excluding Saint Peter the denier, and Judas the betrayer). The five anthers, or male parts of the plant, signify the five wounds, while the three stigmas, or female parts, represent the nails used to seal Christ to the cross. The filaments, or thin stems that support the anthers, radiate from the corona and symbolize a crown of thorns. Swedish botanist Carl Linnaeus, the father of modern taxonomy, would later name the flower for this theological translation.

Even more interesting than the origin of its name, *corona de Cristo*'s sticky, feathery hairs envelop its young fruits and snag uninvited pests in their grasp. The plant hairs secrete chemicals that break down the insects and allow it to absorb nutrients from its victims. This raises an interesting question: Should stinking passionflower / *corona de Cristo* join the Venus flytrap as a distinguished carnivorous plant? The science, as they say, is in progress.[2]

If the above doesn't move you to add some kind of passionflower to your landscape, then note that the plant also plays host to the Gulf fritillary, variegated fritillary, Mexican silverspot, and Julia and heliconian butterflies.

In Texas, the Gulf fritillary is a common visitor from spring through fall and feasts voraciously on the rapid growth of various passionflower species. The spiky orange-and-black caterpillars prefer the underside of the three-lobed leaves, presumably to escape the heat and the radar of bird predators. They are harmless to the touch, by the way.

Females lay yellow-ribbed eggs singly on host plants. They hatch in four to eight days. After about two weeks as a caterpillar, the fritillary morphs into a slender, curved pupa that can range in color from greenish brown to brown with various spots of gray or brown—a great disguise as a dead leaf.

The pupa phase can last five to ten days, depending on heat and humidity. Adults live two to four weeks.

**Plant type: Perennial vine.**
**Light: Sun to partial sun.**
**Water: Low to medium.**
**Soil: Sandy, loamy, calcareous.**
**Size: Can be gregarious and form mats. Climbs trellises, walls, and fences. Will also creep and cover other plants, depending on species.**
**Bloom color: Purple, yellow, green.**
**Fruit: Bright magenta or green, depending on species, the size of a Ping-Pong ball or larger, depending on species. Edible.**
**Bloom time: May to October.**
**Nuisance: Can be rambunctious but is easy to remove. Gregarious growth can form mats over other plants. Climbs trellises, walls, and fences. Choose the species native to your area or most appropriate for your situation.**
**Availability: Seeds widely available, and nurseries often carry plugs or potted plants. Choose your species carefully.**

The Gulf fritillary caterpillar favors passionflower as a host plant.

Don't forget that the fruits of passionflower are edible. They range from green and firm to red and balloon-like. The ones on the fringed passion-flower in my yard are mostly air and serve well as a garnish on salads or an addition to smoothies. The taste is subtle and slightly sweet, with gelatinous matter surrounding the high-fiber seeds.

Passionflower has a long history of medicinal and edible uses. Research shows that extracts from Maypop passionflower (*Passiflora incarnata*) can induce sleep. Stinking passionflower contains extracts that can be useful in treating ulcers.

CAUTION!

The leaves of of *Passiflora caerulea*, known as blue passionflower, can be toxic.

# PEPPERGRASS

*Lepidium virginicum*

Brassicaceae / Mustard family
Culinary / Medicinal / Wildlife

From the wilds of Mexico to empty lots in the United States, *Lepidium virginicum*, known as peppergrass, pepperweed, or "poor man's pepper," finds a home in areas where the soil has been disturbed by human activity like excavation, digging, or construction. It pops up in the "inferno strip" between the sidewalk and the curb, graces alleyways and parking lot perimeters, and self-seeds in exposed pockets of earth or cracks in the sidewalk. As an annual or biennial, depending on weather and conditions, it ranks as a "pioneer plant"—among the first to show up when soil is disturbed, a trait that makes it easy to grow.

Its omnipresence has given peppergrass an unfortunate reputation as an undesirable, disposable weed. But this native of the Brassicaceae, or mustard family, stands out in the Eurasian-dominated field of cruciferous

Peppergrass is known as a "pioneer plant."

Peppergrass seizes opportunity wherever it can—including sidewalks.

vegetables. While its broccoli sisters, cabbage cousins, and mustard brothers have all been cultivated, hybridized, and prized for their agricultural value, peppergrass remains relatively unknown and underappreciated.

Peppergrass starts out short and stout in stature, its small cleft leaves measuring a maximum of three inches. Among the first plants to appear in the spring, it will bolt, reaching a foot tall or more. Eventually, its bristlelike stems generate racemes, thin stalks with discrete flowers featuring four white petals and four green sepals, usually smaller than an eighth of an inch. Flat, oval seedpods with a small notch at the tip come next. These spicy seedpods should be harvested when they're green. Take two fingers and strip them from the stem, from the outside in. Doing so makes a quiet "zip" sound, which seems appropriate given the zippy taste of peppergrass.

Peppergrass works overtime as a caterpillar food for several species of butterflies and moths in their larval stages, including the elegant spotted peppergrass moth (*Eustixia pupula*) and the checkered white (*Pontia protodice*). Apart from *Homo sapiens*, mammals have little interest in the mustardy flavor of peppergrass, although rabbits and groundhogs have been observed eating young plants in early spring when not much else is available.

Plant peppergrass and you're likely to get a checkered white butterfly laying gold, dome-shaped, grooved eggs on the seeds and flowers of the plant. The eggs hatch and morph into handsome gray and yellow caterpillars with black dots, before transforming into interesting bluish-gray

The leaflike seedpods of peppergrass make a peppery condiment.

Peppergrass pods, ready to eat or add to a recipe.

chrysalises with lateral white stripes and small black dots on the ventral half of the body. The pupae attach themselves to a safe place with a cremaster, a sturdy silk button caterpillars spin to support the weight of the chrysalis while it awaits eclosure. It also makes a silk girdle, a thin thread that wraps around the chrysalis and holds it upright, like a harness.

After about ten days, the checkered white will drop from the chrysalis shell. An understated white butterfly with charcoal-colored dots, the species is sometimes considered an agricultural pest. The checkered white serves as an ecosystem pollinator and has been observed nectaring on at least fifty flower species.

If you can't find peppergrass plants in a vacant lot or disturbed area from which to harvest seeds, order them online or find them at a specialty nursery. When ordering, be sure to get Virginia peppergrass, *Lepidium virginicum*. Follow the directions for any mustard plant—plant about three weeks before last frost date.

**Plant type:** Annual herb.
**Light:** Full sun.
**Water:** Drought tolerant, needs little water.
**Soil:** Grows anywhere, in sidewalk cracks, empty lots. Not picky about soil.
**Size:** 1–3 feet tall, 1–2 feet wide.
**Bloom color:** Tiny white flowers.
**Bloom time:** Spring.
**Availability:** Seeds available online. Can also transplant from plants you find in your area in the spring and let them go to seed in your landscape for next year's crop.

Peppergrass seeds and stems smack of mature arugula or peppery watercress. Roots can be ground and combined with salt and vinegar for a horseradish-style condiment. Billed as a "wild spice," peppergrass adds flavor to any salad and spices up pasta sauces, salsa, soups, and stews.

*Mother Earth News* published a delicious chermoula recipe in 2017 that put peppergrass on the culinary radar. It calls for olive oil, salt, garlic, cilantro, a hot pepper, and a tablespoon of fresh green peppergrass seedpod disks. Blend together and slather on seafood, grains, or steamed vegetables.

Peppergrass flaunts understated white blooms.

Not only does peppergrass rank as an overlooked wild spice, but every part of the plant contains antiprotozoal agents—that is, chemical substances that combat infection by single-celled parasitic organisms. The plant has been used for centuries to treat diarrhea, dysentery, liver issues, worm infections, and stomach problems. Native Americans used the bruised fresh plant, and tea brewed from its leaves to treat poison ivy and scurvy. The Houma people of Louisiana and Mississippi made a compound decoction of peppergrass with whiskey to alleviate tuberculosis.

Note: every part of this plant is edible. You can throw the seeds or leaves in salads for a peppery spike like arugula or sprinkle the seeds on salmon or veggies for a pungent kick.

Some people grind up the roots and make a horseradish paste. The recipe below uses peppergrass to make a pesto-like sauce that can be used as a dip or topping on bread, fish, or vegetables or added to salad dressing.

Peppergrass chermoula, ready to serve.

Venison tenderloin dressed with peppergrass chermoula.

## PEPPERGRASS CHERMOULA

*Adapted from* Mother Earth News[1]

*Ingredients*

> 2 large garlic cloves
> ¼ cup fresh green peppergrass seedpod disks
> ½ cup cilantro and/or parsley
> ½ small chile pepper, chopped fine
> ¼–½ cup extra-virgin olive oil
> ½ teaspoon salt
> Fresh-ground pepper to taste

*Directions*

> 1. Place the garlic, peppergrass, chile pepper, and cilantro/parsley in a food processor and pulse to finely chop. Scrape down the sides of the food processor bowl with a spatula and pulse again (repeat a few times to end up with a more or less evenly minced mixture). Alternatively, finely chop the garlic, chile, and cilantro/parsley. Pound them together with the peppergrass with a mortar and pestle.
> 2. Add the salt and ¼ cup of the olive oil and blend. You want to have a slightly liquid paste. Add more olive oil if needed. Add black pepper to taste.
> 3. Store chermoula in the refrigerator for up to 2 months.

# PRICKLY PEAR

### *Opuntia spp.*

Cactaceae / Cactus family
Culinary / Interesting history / Medicinal / Wildlife

The pervasive cactus known as prickly pear checks a lot of boxes: evergreen, drought tolerant, few diseases or pests, and distinctly beautiful. In the spring, its fleshy, oval pads push out beautiful yellow, orange, or bright pink flowers. Edible magenta fruits follow shortly.

The flowers draw bees, beetles, and other insects like a magnet. The cactus wren uses its thorn-covered structure as a shelter.

So many benefits, and prickly pear asks for little in return. With well-drained soil, occasional water, and lots of sunshine, this cactus will thrive in much of the United States.

Myriad insects gravitate to prickly pear.

Prickly pear can create a defensive wall in your landscape. Photo courtesy of Nicolas Rivard.

Planted as a border, in the background, or as a focal point or accent, prickly pear can be used defensively in the landscape. Strategic placement of prickly pear in your yard can keep deer from your vegetable garden or people from getting too close to your bedroom window. Some of us put a prickly pear pad in front of a recently planted seedling to keep dogs from digging or stepping on the new plant. *Opuntia* plants near the vegetable garden should also increase the presence of insect pollinators, which are likely to boost the production of tomatoes, cucumbers, squash, and other summer veggies, since prickly pear flowers are extremely attractive to bees.

With more than ninety *Opuntia* species in the United States, prickly pears come in a variety of heights and flower colors, with varying cold tolerance and different pad sizes and shapes. Some species can be grown in pots. All prickly pears have thorns except *Opuntia ellisiana*, spineless prickly pear, sometimes called tiger tongue or South Texas prickly pear.

While the spineless version doesn't sport the defensive thorns of other *Opuntia* species, it does carry the annoying glochids. These petite spears can get lodged under the skin and cause mild irritation for weeks, so thick gloves and tongs are advised when processing prickly pear.

Cattle ranchers have never appreciated prickly pear when it dominates a field meant for feeding livestock. Left to its own devices in hospitable conditions, prickly pear can spread, sometimes creating undesirable monocultures. Every pad that touches the ground has the capacity to sprout roots and form a new plant. This could be a plus if you are in the prickly pear business, but on an overgrazed prairie, it can present a problem.

During periods of drought, ranchers appreciate prickly pear. That's when they whip out their blowtorches and burn off the plants' thorns and glochids, converting the ornery succulents to cattle feed. This emergency technique, known as *chamuscando*, dates to the seventeenth century and has kept many ranchers afloat during tough times.[1]

**Plant type: Cactus/succulent.**
**Light: Sun.**
**Water: Low.**
**Soil: Not picky. Needs good drainage.**
**Size: Can get 6–9 feet tall, 3–6 feet wide. Forms colonies.**
**Bloom color: Yellow, purple, orange, depending on species.**
**Fruit: Bright red *nopal* fruits, the size of a petite pear, are edible, but be sure to remove spines and glochids.**
**Bloom time: Spring to early summer.**
**Availability: Widely available, easy to grow from a single prickly pear pad if you're patient.**

Bee diving into a prickly pear blossom.

Another unique aspect of prickly pear: an insect called the cochineal scale feeds exclusively on it. A relative of the soft-bodied aphid, this parasite absorbs moisture and nutrients from prickly pear, manufacturing carminic acid in the process. Cochineal also produce a fuzzy white camouflage on prickly pear pads. This sticky substance conceals the squishy female cochineal, which produces the magenta-colored carminic acid. When crushed, the white fuzz disappears into the harmless bright red liquid, staining your finger with a purplish-pink hue.

Carminic acid has been used as a dye for centuries. When Cortés arrived in the New World in the early 1500s, the conspicuous red dye used by the Aztecs caught his eye, and he took cochineal back to Europe.[2] Cochineal is the source of a particular crimson dye used to paint missions, especially in San Antonio.[3]

Cooks and distillers have also tapped the rich red to make food and drink more appealing. The deep crimson of the Italian aperitif Campari was made possible for more than a century by the addition of cochineal. The dye is also used in cakes, pastries, and yogurt as well as in lipstick, blush, and other cosmetics.[4]

About seventy thousand of the dried insects are required to make one pound of cochineal dye, which is FDA approved, tasteless, and noted on packaging as carminic acid or cochineal.

For centuries, prickly pear pads—actually flattened stems—have been utilized as food. After the spines and tiny barbed glochids are removed, young pads are boiled and consumed as *nopalitos*. The high-fiber vegetable tastes like green beans or okra. It can be added to scrambled eggs to make breakfast tacos, breaded with corn flour and fried, or added to stews and soups, which it will thicken because of its gelatinous properties, much like okra. The eye-catching fruits, called tunas, can be thrown into a smoothie, used in a fruit-flavored refreshment called an *agua fresca*, or prepped as an addition to cocktails.

The medicinal applications of prickly pear throughout history include use as an analgesic to relieve back pain, as a burn dressing, for rattlesnake bites, wart removal, to stimulate the flow of breast milk, to encourage urination, to alleviate constipation, and much more. The thorns were tapped to pierce ears and make tattoos.

More recently, prickly pear has been studied for its ability to control blood sugar levels in diabetics. The mucilaginous properties it shares with okra slow down dietary sugar uptake.[5] A 2010 survey found that prickly pear has great potential as a method for filtering water. According to the research, prickly pear gel filtered out 98 percent of bacteria in a contaminated water sample. Researchers suggest that the cactus could become a sustainable and affordable water purification method in developing countries.[6]

Prickly pear fruit can be tapped for smoothies, paletas, or shrubs.

# RED YUCCA

### *Hesperaloe parviflora*

Asparagaceae / Asparagus family
Wildlife / Xeric / Edible flowers

A popular landscape plant worldwide, the evergreen red yucca thrives in a variety of soils, requires little water, and proves irresistible to pollinators of all kinds. Its reddish-pink flowers shoot skyward in spring and last for weeks. The plant is pretty much maintenance-free and can be used as a border, an accent, or a centerpiece. It even grows in pots.

It's not hard to understand why red yucca ended up in the asparagus family.

Known as flowering red yucca, red yucca, coral yucca, hummingbird yucca, samandoque, and false yucca, this plant has many names, but the latter name may be the most apt. Why? Because *Hesperaloe parviflora* is not a yucca at all.

Taxonomically, red yucca began in the agave family because of its physical resemblance to yuccas and agaves. It was first collected by Charles Wright from gravelly hills near the mouth of the Pecos River in Val Verde County, Texas, and placed in the yucca family in 1859.[1] Over time, it was reclassified in the aloe family and eventually the asparagus family. Now several species of *Hesperaloe* are recognized.

The flowers that emerge from the panicles, or spikes, of this widely adopted xeriscape plant in the spring can flourish until fall. They make the plant resemble a rose-colored asparagus stalk before they fully open. The Latin name *Hesperaloe parviflora* translates as "fake aloe with small flowers."

Red yucca flowers are edible for people, but don't let your dog nosh on them. Red yucca contains saponins, a bitter-tasting glycoside that causes foaming and can induce vomiting in canines. "Saponin" comes from the Latin *sapo*, which means "soap."

Low-maintenance red yucca has become a popular landscape choice in recent years.

Red yucca sports delicate flowers that are edible for humans, but not dogs.

As with many "poisonous" plants found in the landscape (milkweed, Jimson weed, and lantana are other examples), a bitter taste produced by toxic chemicals discourages consumption, so dogs and others are turned off by the nasty flavor. Generally, canines just taste red yucca and rarely eat enough to cause serious consequences.

That said, discourage your dog from chewing on *Hesperaloe parviflora*. None of my dogs have ever been attracted to it, but they have sought out the lantana in my yard, eating it as a seeming stomach cleanser. Beyond a mild purging—usually a small amount of bile coughed up accompanied by a wad of chewed-up plant matter—no great harm has ever resulted.

A native of the Chihuahuan Desert from New Mexico and Texas south into Mexico, red yucca flaunts a fleshy, almost bluish-green rosette at its base. This base supports towering flower stalks that grow from its center and can reach a height of five feet. Another appealing feature: generally, deer avoid the foliage, but they may eat the blooms. And as an evergreen, red yucca provides welcome greenery year-round and requires practically zero care. Nurseries market it as "maintenance free."

Red yucca seedpod.

With their yellow- and peach-colored sepals and brighter red external petals, the bell-shaped tubular blossoms are a favorite of butterflies, bees, and especially hummingbirds. In the evening, moths visit the inviting blossoms.

The plant also develops chubby three-chambered seedpods in late summer. Dramatic flat black seeds shaped like the letter *D* rest inside the spheres, which start out green and then dry out and turn brown. If you intend to plant the seeds, be sure to harvest only from completely dry, open pods. The seeds can remain viable for years when refrigerated.

Red yucca is an excellent choice for hot, dry areas and poor, well-drained soil. Flowers bloom all summer. It has no diseases or insect predators and requires little attention other than removing the dried, spent flower stalks each winter.

Plant type: Perennial succulent.
Light: Prefers full sun but can handle some shade.
Water: Extremely drought tolerant.
Soil: Well drained.
Size: 2–4 feet tall, but flower spike can reach 5–6 feet; 2–3 feet wide.

**Bloom color: Coral, red, pink.**
**Fruit: 3-chambered seedpods form in late summer. Not edible.**
**Bloom time: Spring.**
**Availability: Widely available at nurseries as potted plants. A new hybrid with yellow flowers is also available.**

Red yucca flowers are edible. To add a flash of color to a salad, toss it with young raw flower buds. Or pull the reddish flowers off the stalk, cut off the base, and sauté with onions. Dress up an omelet[2] or add the cooked flowers to grains.

*Hesperaloe parviflora* was reportedly used as a soap because of the natural saponin in its roots. Indigenous people made tea from the pounded dried roots to treat kidney problems. Some would chew the roots as a cure for intestinal parasites or diarrhea.

If you're feeling crafty, split red yucca leaves can be woven into baskets.[3] The Pame tribe near San Luis Potosí used a different species of *Hesperaloe* in a mix with maguey to craft rope. The tiny threads that curl off the plant's fleshy leaves look like a natural floss, and the plant has been utilized to make paper.

# SAGE

*Salvia spp.*

Lamiaceae / Mint family
Culinary / Medicinal / Wildlife / Spiritual

---

Here's some sage advice for those wanting to make the most of their yards: grow *Salvia*. With about nine hundred species, a sage most likely exists to match your color palette and ecoregion. White, blue, purple, and scarlet flowers grace this most numerous member of the mint family, classified as *Salvia*. Originating in the Mediterranean, common sage and other *Salvia* species do well in most soils with good drainage, tolerate heat, and don't appreciate overwatering.

Honeybee on sage.

They can be evergreen or deciduous, and perennial, biennial, or annual. These impressive shrubs and flowering plants range in size from eighteen inches to five feet tall, depending on the species. Their gray-green foliage gives off a pungent and pleasant fragrance that repels garden pests and deer but serves well in the kitchen. Plus, sages are drought tolerant and easy to grow.

*Salvia* species also pack a hefty pollinator appeal.

Butterflies, bees, and hummingbirds linger on the tubular flowers of *Salvia*. And the foliage serves as a host plant to *Lintneria eremitoides*, the sage sphinx moth, a relatively rare moth native from the Great Plains south to Texas.

Sages exhibit an unusual flower structure. The stamen, the male part of the flower that contains pollen, sports a "lever" known as the staminal lever mechanism. As pollinators seek out nectar in the cup of the flower, they brush against this lever, releasing pollen—which conveniently lands on the pollinator, practically ensuring that it will find its way to the next plant.

The name "sage" originates from the Latin *salvare*, which means "to save" or "to cure." During medieval times, it was known as "salvation plant" and was a common fixture in monastic gardens.[1] The use of "sage" as an adjective or a noun to describe someone who is wise comes from its power to strengthen memory, for which it was used in ancient medicine.[2] Its memory-enhancing compounds have since been scientifically proven and demonstrate cognitive-enhancing effects in human adults.[3]

Technically, all sages are nontoxic and thus edible—but that doesn't mean they all taste good. In my view, none can be described as delicious directly from the plant.

My personal favorite, mealy blue sage (*Salvia farinacea*), a native Texan, is not especially tasty, but it loves calciferous soils and distinguishes itself with velvety purple flowers projecting from upright square stems—another trait of *Salvia*. The plant makes a striking background or accent with its projectiles of blue velvet blooms that are a magnet for butterflies, bumblebees, and other pollinators. Goldfinches feast on its seeds. And mealy blue is amazingly resilient. Our dogs have trampled one bush in the front yard several times, and the plant rebounds with vigor.

Sage plays well with others.

**Plant type:** Perennial herb.
**Light:** Sun. Some species can handle partial shade.
**Water:** Low.
**Soil:** Well drained. Does not like clay.
**Size:** 2–3 feet tall, 1–3 feet wide.
**Bloom color:** Purple, blue. Various species offer red, white, and other colors.
**Bloom time:** Spring to late fall.
**Availability:** Widely available in seeds and plugs.

Even common sage, *Salvia officinalis*, considered the culinary *Salvia* of choice and a mainstay of any herb garden, has a minty but off-putting taste and texture when eaten raw. That said, fresh sage leaves have been chewed for centuries as a breath freshener and teeth cleanser.[4] These days, we frequently use it dried as a seasoning, but when the fresh, fuzzy leaves are sautéed crisp and sprinkled with salt they assume a savory, almost nutty flavor that begs the question "What *is* this?" Eat them straight, deck them atop your pasta, or crumble into pestos, chutneys, and cheeses for a delightful, surprising crunch.

Dried or fresh leaves can be used to season meats, stuffings, or stews, or to make sage butter. Common sage flowers are also edible.

Other sages also provide culinary benefits. Pineapple sage (*Salvia elegans*), for example, native to Mexico and Guatemala, blooms red and thrives in hot, dry climates like those in the Southwest. According to the Lady Bird Johnson Wildflower Center's "Ask Mr. Smarty Plants," any red *Salvia* flower tastes good, and those with warm-colored flowers will be sweet. Pineapple sage appears to fit this mold. Its flowers have a sugary note, and the leaves smack of pineapple.

Sage has been tapped as a medicinal herb since ancient times. Pedanius Dioscorides, a military physician in the Roman army and author of the first bible of pharmacopeia knowledge, *De materia medica*, viewed sage as one of the most important herbs of the time.[5] He used it as a decoction (a reduction of the plant made by boiling or stewing to extract its compounds), on wounds to stop bleeding, and as an ulcer treatment. He also utilized it as a tea for sore throats and hoarseness.

Ancient Egyptians used sage to address infertility. The Mazatec shamans of Oaxaca, Mexico, continue to use *Salvia divinorum*, known as seer's sage, in religious rituals. Native Americans rolled the leaves of *Salvia apiana*, or white sage, into a bundle and lit the foliage in a tradition accompanied by prayer known as "smudging." The pleasant scent of the sage smoke was believed to carry messages to the gods, as well as clear the space of negativity.

A 2017 study explored in depth the particular medicinal attributes of common sage. Sage compounds sound like a cure-all, demonstrating anticancer, anti-inflammatory, antioxidant, antimicrobial, hypoglycemic, memory-enhancing, and other effects. In short, we should all be eating more sage and drinking more sage tea.

### Make a Sage Smudge Stick

Cut small branches and leaves from the plant, fold up into a bundle, and tie with string or twine. Hang in a cool, dark place and let it dry out for about a week. (I like to put it over the rearview mirror in my car for good luck and great scent.) When ready to use, light a match and smudge your space.

## CRISPY FRIED SAGE LEAVES

*This is so easy to make, I hesitate to call it a "recipe." Just fry common sage leaves in oil and sprinkle with salt. You'll love it.*

*Ingredients*

    About ¼ cup (or less) olive oil—enough to coat your sauté pan with about 1/16 inch of oil

    Fresh common sage leaves

    Salt

*Directions*

    1. Gather sage leaves and rinse. Pat dry with a clean towel.

    2. While they're drying, heat oil in pan.

    3. When oil is hot, add leaves and fry, about 1–2 minutes. Flip them with tongs and fry the other side.

    4. Remove with spatula or slotted spoon and lay on paper towels. Sprinkle with salt immediately, while still hot, so the salt sticks.

    5. Eat as a snack, add to a snack mix, or put on top of pasta, cheeses, or rice—it will add a crispy, salty, earthy flavor.

*Note: these are best fresh out of the pan.*

This common sage, planted at a ranch, thrives and provides herbs for cooking, despite never receiving supplemental water.

# SUNFLOWER

### *Helianthus spp.*

Asteraceae / Aster family
Culinary / Great cut flowers / Heliotropic / Medicinal / Wildlife

Easy-to-grow sunflowers amaze with their generosity. Depending on which of the more than seventy species of *Helianthus* you choose, one tiny seed can result in hundreds, even thousands of tiny florets on the head of the plant.

Yes, that's correct. *Each* of those fluffy yellow growths extending from the center of the sunflower pictured here constitutes an individual flower.

For butterflies, this represents a nectar cafeteria. They can land in one spot and conveniently slurp on serial nectar straws without even changing position. Each flower later turns into a seed that birds and people seek and consume.

Bee on sunflower.

Yes, each of these protrusions is a separate flower that will later form a seed.

Start sunflower seeds inside as early as January and transplant the seedlings to the garden in early spring, or after the danger of first frost is past. You can also sow them directly in the soil, depending on your location and the variety of sunflower chosen.

Soon, the stalky giants of some varieties like mammoth sunflower will reach twelve feet tall, their broad, foot-wide faces perching in the front yard like soldiers offering a welcome salute. They lose their perky dispositions as summer takes hold. Their heads drop and seeds form in place of the flowers.

Leave them in place and birds will help themselves. Several butterfly species will use certain sunflower species as a host plant for their caterpillars. When the plant matures and the seed head dries, the seeds disperse to the ground, where fowl, squirrels, and other critters gather them for a handy protein pop. One tablespoon of sunflower seeds contains four and a half grams of protein, and a single sunflower head can hold between one thousand and two thousand seeds.[1]

If ample flowers, prodigious seeds, and benefits to wildlife don't merit your consideration of *Helianthus* species, how about heliotropism? Sunflowers exhibit the endearing botanical trait of tracking the sun. The flowers, in their youth, literally follow the sun as it moves across the sky.

This has intrigued scientists and others for centuries. In a recent paper, researchers determined that the sunflower's movement is caused by its particular circadian rhythms. Young flowers face east as the sun comes up and slowly follow it west as it moves across the sky. At night, the flowers reset, gradually turning east again to begin the cycle anew. Technically, what we call "turning" or "tracking" the sun is a result of different sides of the plant elongating at different times of day in response to light.

Mature sunflowers don't do this. Typically they live out their lives permanently facing east. Scientists forced some mature flowers to face west and learned that those allowed to face east attracted five times as many pollinators as the west-facing ones. Warmth attracts insects.

At the end of their lives, sunflowers typically drop their heads. When their back sides turn brown, the seeds are ready.

While the mammoth sunflower is a hybrid, native species are widely available and deserve attention. The common sunflower, *Helianthus annuus*, is the most common and provides prime habitat for dove and quail.

Giant swallowtail takes a break on cowpen daisy, another member of the sunflower/aster family.

Its gangly growth habit sometimes doesn't fit home garden landscape plans, but for wildscaping or restoring habitat, it serves well.

Many sunflower cultivars—that is, cultivated varieties—have been developed. As mentioned earlier, while these offer special beauty, the inbreeding and hybridization that accompany cultivated varieties often require trading ecosystem benefits for appearances.

For example, certain cultivars grown for use in the floral and cut-flower industries boast of a "pollenless" head. These varieties were developed from sterile male plants and produce no reproductive pollen. They don't cause allergies, nor do they appeal to bees.

Sunflowers have been grown as an agricultural crop for millennia. During the last three thousand years, indigenous peoples have changed the genetic makeup of this productive native, increasing its seed size by 1,000 percent by selecting the largest seeds for planting.[2] According to industry reports, the global sunflower industry—seeds, oil, and flowers—is on a roll and predicted to grow from $20.5 billion in 2022 to $30.52 billion by 2032.[3]

Plant type: Annual herb.
Light: Sun.
Water: Low to medium.
Soil: Dry clays to rich loams, depending on species.
Size: 4–12 feet tall, depending on species.
Bloom color: Yellow, but cultivars have been developed in almost every color.
Bloom time: Summer to fall.
Availability: Widely available as seed, sometimes available in pots.

When you remove the flowers, the immature seeds cover the head and taste like corn.

The flowers, seeds, sap, and oil of *Helianthus* have also been used for thousands of years by native peoples. Teas, infusions, and powdered forms of the plant's parts were utilized to address chest pain, pulmonary discomfort, and fevers.

Eat the seeds raw or toasted (with or without hulls), add them to a trail mix or salad, use them to fill your bird feeder, or save for next year's crop. Seven mammoth sunflowers I planted in my front yard one spring yielded one and a quarter pounds of seeds.

### GRILLED SUNFLOWER HEAD

Heirloom seed company Baker Creek created a stir in the summer of 2020 when its on-site chef Jenna Asher posted a recipe and video for how to make a grilled sunflower head on the company's Instagram feed. "Like corn on the cob!" read the subhead in the company's 2020 catalog.

According to Baker Creek, grilling a sunflower head is not new. Apteka, a vegan restaurant in Pittsburgh, Pennsylvania, has been offering roasted whole sunflower heads for years. We tried this recipe several times, and it *does* taste like corn on the cob.

*Tip: make sure the sunflower head is not overripe, meaning it hasn't yet formed seed hulls around the kernels. Our first try at this was a disaster and tasted like chicken scratch.*

*Ingredients*

    1 large immature sunflower head

    1 jar sun-dried tomatoes

    Handful of fresh basil leaves, cut in thin strips

*Directions*

    1. Pick the sunflower when the seeds have formed but the shells are still soft. The seeds should have the texture of corn kernels.

    2. Remove all flowers.

    3. Coat sunflower head with oil of your choosing; Baker Creek recommends the oil from sun-dried tomatoes.

    4. Put the sunflower on the grill face down, over medium heat. Cover and cook for 5 minutes.

    5. While sunflower is grilling, finely chop the sun-dried tomatoes.

    6. Remove sunflower head from grill, slather with sun-dried tomatoes, salt and pepper to taste, and sprinkle with basil.

    7. Eat like corn on the cob or use a fork to separate the cooked seeds from the flower head.

*Note: you can use a variety of spices here, not just sun-dried tomatoes— garlic, salt and pepper, pesto, anything that tastes good with corn would work.*

Grilled sunflower head, before dressings are applied.

Grilled sunflower head with sun-dried tomato jam.

# TURK'S CAP

### *Malvaviscus arboreus var. drummondii*

Malvaceae / Mallow family
Culinary / Grows in shade / Medicinal / Wildlife

Famous for thriving and blooming in shade, versatile Turk's cap provides myriad benefits in various habitats.

Known as Texas mallow, sleeping hibiscus, and forest rose, Turk's cap shares the trait of conspicuous blooms with its fellow members of the mallow family—hibiscus, okra, and cotton—but it keeps its flowers tucked into a compact form.

Turk's cap flower and immature fruit.

Turk's cap fruit.

When mature, Turk's cap boasts an impressive staminate (male flower part) column, similar to that of hibiscus, that juts from the center of the closed red flower folds. In the center of this pinwheel, a marble-sized fruit, sometimes called a "Mexican apple," develops in late summer.

Turk's cap has adapted to a wide variety of soils and moisture conditions. In the wild, it typically grows along streambeds in well-drained soil. The drought-tolerant multibranched bush can reach four to five feet in height and tolerates extreme summer heat. It does well under trees or in the sun. Coddle this plant and it might need a trim, especially without the occasional freeze.

Turk's cap seeds provide protein and starch, the leaves contain minerals, and the fruit is rich in vitamin C. Fresh flowers contain a tasty dose of nectar, which draws hummingbirds and butterflies.

The leaves of Turk's cap serve as host fodder for at least three butterfly species that have coevolved with Texas mallow: the Turk's-cap white-skipper (*Heliopetes macaira*), the mallow scrub-hairstreak (*Strymon istapa*), and the glassy-winged skipper (*Xenophanes tryxus*). Its red flowers are a favorite of the cloudless sulphur, a common yellow butterfly that favors open spaces.[1]

The Turk's-cap white-skipper has a delicate appearance, mostly white with tawny brown markings on its outer wings and the edges of its underwings. Its wingspan reaches an inch to an inch and a half. Its caterpillar, in later stages, looks like a fuzzy white worm with a smooth brown head.

Turk's cap thrives in shade.

In its final instar, the caterpillar folds itself into a leaf tent and turns into a green pupa before morphing into a butterfly. The range of the Turk's-cap white-skipper extends from South Texas down to Central America.

And why the odd name "Turk's cap?" The compact red bloom resembles (kind of) a fez, the traditional red felt hat with a complicated history. Conical in shape and typically sporting a tassel, the fez lost favor in the 1800s when it became associated with the oppressive Ottoman regime of Turkey. The similarity of the iconic hat to the bloom of *Malvaviscus arboreus* var. *drummondii* can presumably be tied to the traditional red color of the original fez. Traditional red fezzes, originating in Fez, Morocco, were dyed using the Cornelian cherry, a plant found only in that area of North Africa.

The *drummondii* in the Latin name of Turk's cap comes from the intrepid Irish botanist Thomas Drummond, who collected thousands of specimens in the 1830s and first identified the plant. Drummond's collection was the first from Texas to be widely distributed around the world, and many Texas plant species are named in his honor.

**Plant type: Perennial shrub.**
**Light: Part to full shade.**
**Water: Medium.**
**Soil: Very tolerant. Grows in sandy, loamy clays and limestone soils. As a shade plant, it likes to be under trees.**
**Size: 2–9 feet tall, 3–6 feet wide.**
**Bloom color: Red.**
**Fruit: Dark red, round, 1 inch across.**
**Bloom time: May to November.**
**Availability: Widely available at nurseries in containers.**

The leaves, flowers, and fruits of Turk's cap are all edible. Foragers rave about the plant's many health benefits, especially the antioxidants stored in its flowers. I often pop a fresh flower or fruit in my mouth straight from the plant. The flowers have a subtle sweetness and rose flavor as well as an interesting texture, while the fruit carries some of the same notes, punctuated with the crunchy seeds.

For a tasty, soothing tea, crush up a handful of Turk's cap fruit, steep in hot (but not boiling) water for about fifteen minutes, add honey, and enjoy.

## TURK'S CAP PANCAKES

*Note: you can use a pancake mix if you like, but pancakes from scratch are not difficult, and the ingredients are typically on hand in the average kitchen. I adapted the recipe below from Mark Bittman's, then dressed it up based on a trip my family took to Real de Catorce, Mexico, one summer. We abbreviated the name to R14 since* catorce *means "fourteen" in Spanish.*

There, in a charming old mining town loaded with history, we had these amazing pancakes embedded with local fruits—banana, mango, papaya. The fruits were laid on top of the pancake dough in a hot pan before the pancakes were flipped. This resulted in rich, caramelized fruits—and a delicious and whimsical breakfast. Our young boys were smitten with "R14 Pancakes," and we've been making and serving them ever since. The addition of Turk's cap flowers is our latest innovation.

Deck with a pat of butter and a dribble of maple syrup. Yum.

Turk's cap flowers.

Turk's cap pancakes, ready for the pan.    Breakfast is ready!

*Ingredients*

   2 cups all-purpose flour

   2 teaspoons baking powder

   ¼ teaspoon salt

   2 eggs

   1½ to 2 cups milk

   2 tablespoons melted and cooled butter (optional), plus unmelted
   butter for cooking

   ½ cup fresh Turk's cap flowers

*Note: you can also add other fruits like banana, pear, apple—whatever
you have on hand.*

*Directions*

   1. Mix dry ingredients in medium bowl.

   2. In a separate bowl, beat eggs into 1½ cups milk, then stir in 2
   tablespoons melted cooled butter, if using it.

   3. Gently stir this mixture into dry ingredients, mixing only enough
   to moisten flour; don't worry about a few lumps. If batter seems
   thick, add a little more milk.

4. Heat a griddle or large skillet over medium-low heat. Place 1–2 teaspoons of butter or oil in pan. When butter melts and foam subsides, ladle batter onto pan, making pancakes any size you like.

5. While dough is still wet in pan, place Turk's cap flowers in a circle, like the numbers on a clock. Put slice of banana in middle.

6. Adjust heat as needed. Typically, the first batch will require higher heat than subsequent batches. Flip pancakes after bubbles rise to surface and bottoms brown, after 2–4 minutes.

7. Cook until second side is lightly browned and fruit is caramelized.

8. Serve with butter and maple syrup; garnish with a fresh Turk's cap flower.

# WILD ONION

*Allium canadense* **and other** *spp.*

Amaryllidaceae / Amaryllis family
Culinary / Medicinal / Insect repellent / Wildlife / Deer resistant

Myth and science merge in wild onion, also known as wild garlic or meadow garlic. Since ancient times, members of the amaryllis family have been hailed as magical herbs. Found in Greek temples and the Egyptian pyramids, garlic was fed to Olympic athletes as the first performance-enhancing drug. When King Tutankhamen's crypt, dated to 1500 BC, was excavated in 1922, intact cloves were found nestled in the tomb.[1]

Field of wild onion along the Llano River.

The Greeks and Egyptians were onto something. For centuries, folk medicine has used onion and garlic to treat dysentery, increase fertility and productivity, stave off infection, and preserve food. And science has since proven the medical benefits of *Allium*, including as a treatment for high blood pressure. Plus, it tastes good.

The plant is easy to grow in your garden or a container. You can plant organic garlic cloves from the grocery store (as long as they're relatively fresh) or secure them online or from a local nursery. You can also cut an onion in half, put toothpicks in it to suspend it over water, and wait for roots to form. When they do, you can transplant the onion into the soil and make more onions.

But in the spring, wild onion is most likely thriving somewhere near you—probably near a stream or in a watershed. The compounds in onion responsible for its distinctive odor make it deer proof, which helps its establishment. Its bulbous root ball keeps it intact during floods and holds soil in place. It looks like chives on steroids. Go look for it.

In Texas, almost a dozen wild onion species flourish. The most common, *Allium canadense*, is native to almost all of the United States.

Like all *Allium* species, the herbaceous perennial *Allium canadense* boasts flat leaves, hollow stems that support a lovely purple or white flower, and a marble-sized white or purplish bulb underground. Pulling wild onions from moist earth makes a ripping sound, like the tearing of fabric, as the tiny roots separate from mud. If you harvest from the wild, take a spade and dig up a set for placement in a pot or transplanting to your yard.

CAUTION!

Be advised that you should not consume parts of any wild edible plants, herbs, weeds, trees, or bushes until you have verified with a health professional that they are safe. As with any new foods, best practice suggests introducing them slowly into your diet in small amounts.

Whole wild onions, recently harvested.

Wild onion recedes in hot weather but bounces back when rain and cool temperatures return. It can create a carpet of green, and some consider it a nuisance. But for *Allium* lovers, it's a gift.

Wild onion is an easy plant to grow and is not as strong as the store-bought variety. Think of an extremely compact green onion. It does well in a pot on the kitchen windowsill or in a patio container and invites a snipping of its greens to add to scrambled eggs, garnish a salad, or sprinkle on bread with olive oil to toast under the broiler.

The version I have in my downtown garden, *Allium canadense*, was transplanted from the floodplain of our ranch on the Llano River in the Texas Hill Country where our entire family marvels at the stages of its cycle. "Green onions are back!" my sons will text me following a good rain. The *Allium* stand in our watershed flourishes and recedes in a predictable cycle, depending on heat and moisture.

In the garden, the plant is pretty predictable. It appreciates regular water in a pot or in the ground, and it blooms and flourishes in spring to early summer. The flowers then generate tasty seedpods or bulblets that I call "onion bulbs," which we harvest for various culinary uses.

Wild onion bulblets.

Pollinators appreciate *Allium* for its nectar and pollen, but in general, insects don't care to nosh on its foliage, since it radiates a pungent smell.

*Allium* is known for high deer resistance, yet some folks have noted that when wild pigs forage on them, an agreeable onion-like taste accompanies the meat when cooked.

Some gardeners recommend planting onion and *Allium* species near your porch or walkways to deter mosquitoes. Or companion-plant them next to other vegetables to repel predatory insects like aphids and beetles that don't appreciate their smell.

"Society garlic" (*Tulbaghia violacea*), popularized in recent years with commercial growers, hails from South Africa and is technically not an *Allium* but looks, tastes, and smells like one. Like onions and garlic, it's in the amaryllis family and shares the pungent smell. Society garlic's smell can be detected from yards away and is especially strong when brushed against. The "society" moniker intrigues, given that in ancient times, garlic was considered a working-class herb. Its cloves were fed to laborers who built the pyramids, in the belief that it made them stronger and more productive.

While the strong-smelling foliage deters insects and wildlife, *Allium* flowers attract bees. The plant offers both nectar and pollen early in the spring.

With its conspicuous flowers, deer resistance, bee friendliness, and culinary and medicinal uses, multitasking wild onion merits a spot in your landscape.

Plant type: Perennial grass.
Light: Prefers full sun but can tolerate partial shade.
Water: Low to medium.
Soil: Well drained.
Size: 8–12 inches tall, 1 foot wide.
Bloom color: Purple, lavender, white.
Fruit: Small. Tiny white edible bulb.
Bloom time: April to July.
Availability: Seeds and bulbs widely available.

Every part of this plant is edible. The tiny bulbs can be sautéed with vegetables, crushed and smeared on bread and broiled, or studded into a cut of meat. Chop up the stems like chives and throw them in a salad or add to Greek yogurt. The purple or white flowers make a lovely garnish and are completely edible. And when the flowers go to seed, harvest the bulblets for next year or make a pesto, throw onto a pizza, or sauté with olive oil and veggies for a uniquely textured, oniony kick.

Eastern black swallowtail on wild onion bloom.

## SAUTÉED WILD ONION BULBLETS WITH MUSHROOMS

*Ingredients*

    1 tablespoon olive oil

    ¼ cup wild onion bulblets

    2 cups mushrooms

    Salt and pepper to taste

    Parmesan cheese (optional)

*Use as topping on pasta, veggies, pizza, or eggs—as a side dish or main course of your choice.*

*Directions*

    1. Heat olive oil in cast-iron pan.

    2. Add onion bulblets, stir, and sauté for 3 minutes

    3. Add mushrooms, stirring so they don't burn.

    4. Cook to desired consistency (I like mine firm and not soggy).

    5. Salt to taste, add to desired side dish or main course, and enjoy!

Sautéed potatoes and wild onion bulblets.

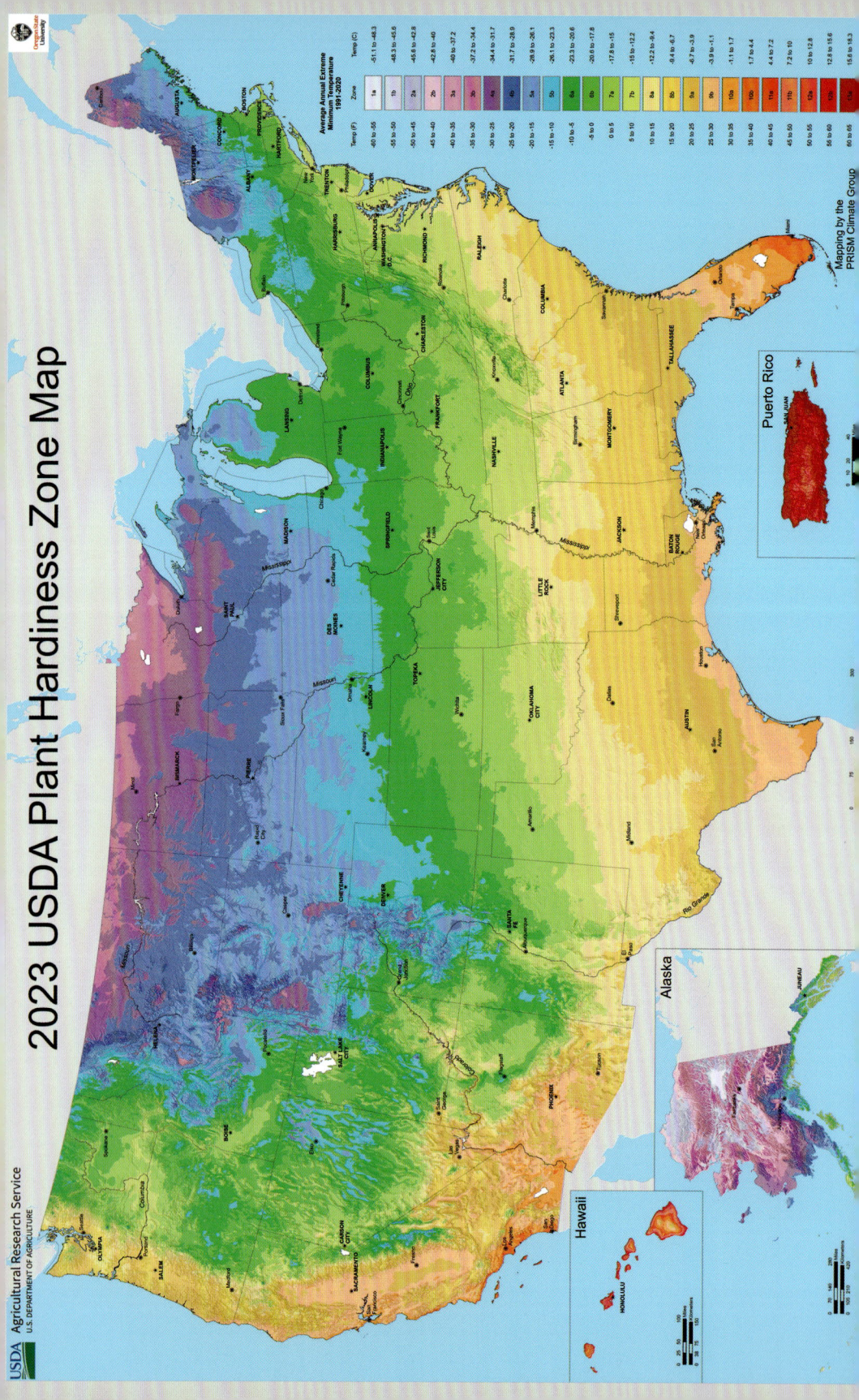

# 2023 USDA Plant Hardiness Zone Map

USDA Agricultural Research Service
U.S. DEPARTMENT OF AGRICULTURE

### Average Annual Extreme Minimum Temperature 1991–2020

| Temp (C) | Zone | Temp (F) |
|---|---|---|
| -51.1 to -48.3 | 1a | -40 to -55 |
| -48.3 to -45.6 | 1b | -55 to -50 |
| -45.6 to -42.8 | 2a | -50 to -45 |
| -42.8 to -40 | 2b | -45 to -40 |
| -40 to -37.2 | 3a | -40 to -35 |
| -37.2 to -34.4 | 3b | -35 to -30 |
| -34.4 to -31.7 | 4a | -30 to -25 |
| -31.7 to -28.9 | 4b | -25 to -20 |
| -28.9 to -26.1 | 5a | -20 to -15 |
| -26.1 to -23.3 | 5b | -15 to -10 |
| -23.3 to -20.6 | 6a | -10 to -5 |
| -20.6 to -17.8 | 6b | -5 to 0 |
| -17.8 to -15 | 7a | 0 to 5 |
| -15 to -12.2 | 7b | 5 to 10 |
| -12.2 to -9.4 | 8a | 10 to 15 |
| -9.4 to -6.7 | 8b | 15 to 20 |
| -6.7 to -3.9 | 9a | 20 to 25 |
| -3.9 to -1.1 | 9b | 25 to 30 |
| -1.1 to 1.7 | 10a | 30 to 35 |
| 1.7 to 4.4 | 10b | 35 to 40 |
| 4.4 to 7.2 | 11a | 40 to 45 |
| 7.2 to 10 | 11b | 45 to 50 |
| 10 to 12.8 | 12a | 50 to 55 |
| 12.8 to 15.6 | 12b | 55 to 60 |
| 15.6 to 18.3 | 13a | 60 to 65 |

Mapping by the PRISM Climate Group

Puerto Rico

Alaska

Hawaii

# FROM THE GROUND UP:
# GUIDANCE ON PLANTING WITH PURPOSE

I've tried to be conversational about gardening and the use of plants recommended in this book. That said, anyone who wants to be successful at growing the plants profiled here will need to learn a bit of gardening jargon. What follows is basic growing guidance as well as a resource section.

First off, don't be intimidated! The plants profiled were specifically chosen because they're relatively easy to grow. Most are natives. The others are desirable nonnatives. A successful plants-with-purpose garden aspires to provide these plants with an environment that mimics nature as much as possible.

### Native, desirable nonnative, invasive—What does it mean?

"Native" means the plant occurs naturally in the place where it lives, without human intervention, help, or transportation. That means the plant evolved where it grows, speaks the language of the local ecosystem, and contributes in some fashion by providing ecosystem services useful to the local community—for example, shelter or food for the wildlife that live there, or erosion control along a streambed.

A nonnative is simply a plant that is not originally from the place where it's currently growing. A desirable nonnative plant is one that is well behaved and acts as a good member of the local community by providing some type of ecosystem service.

Some plants are not well behaved. They consume more than their fair share of water, light, and available soil nutrients, crowding out other plants that live in the neighborhood. Many variables determine behavior—the species, the soil, how much water they absorb, the geographic circumstances of where the plant is planted, and where it originated.

189

When a plant is rambunctious and overzealous, like a dinner guest who talks over everyone or drinks all the wine, that plant is a potential nuisance. Such plants can still supply services to the local critters and ecosystem but they simply are rude in doing so. They need a bit of gardening discipline. As for the plants profiled here, potential overzealous growth is noted as "nuisance potential" and is generally applied to particular species of recommended plants.

Gardening websites and many books bandy about the word "invasive" to describe rambunctious and unruly plants. "Invasive" can be a highly charged term in gardening and horticultural circles. An invasive is a plant that is not native and not well behaved in certain environments. Invasive plants are overly domineering to the point of causing harm. They do not abide by the ground rules of contributing to the community or serving the common interest. Consider them uninvited guests or unwelcome drop-ins. Don't let them make themselves at home.

"Invasive" has a very specific meaning within government agencies responsible for managing federal lands. For example, for the US Forest Service, an "invasive" plant is one that is not native to the ecosystem where it is being considered for planting, and its introduction to that ecosystem causes or is likely to cause economic or environmental harm. In some states, planting or transporting plants designated as invasive species can result in fines of up to $500.

Because of this technical definition, I have mostly avoided using this term in the book because the plants with purpose profiled here are generally considered noninvasive, although you will encounter the generic use of the term if you do some research (lantana). Certain species can definitely become overzealous growers (goldenrod), but that is also in the eye of the garden beholder. I allow my goldenrod to run rampant and serve as living mulch. When I want to plant something else, I pull it out with ease.

Even so, whether or not a plant is invasive depends entirely on the geographic location and climate where it will be planted. Certain species will dominate and cause damage in some parts of the country, yet will behave appropriately elsewhere. As always, do your homework and choose a suitable species.

**Hardiness/grow zones and ecoregions**

Speaking of zones, a basic understanding of hardiness zones and their more useful paradigm, ecoregions, is helpful to gardeners at all levels.

When considering a plant, a common question is, will it survive a freeze? Or, in garden parlance, how hardy is it?

Citrus, for example, is generally not freeze tolerant; thus, it's inadvisable to plant it in zones where temperatures drop below 32° Fahrenheit for more than a few hours. Unless protected on a patio or with heat lamps, your tree will freeze and die.

But how do you know your hardiness zone? Consult the USDA Plant Hardiness Zone Map.

For decades, hardiness zones have been a guide for gardeners to help determine the appropriateness of a plant for a specific region as well as when to plant seeds of a particular plant. The map, developed by the US Department of Agriculture in 1960[1] and most recently updated in 2023, divides the United States and Canada into thirteen zones, based on the average annual minimum winter temperature.

Zone 1 starts up north and represents the coldest areas of the country, while Zone 13 is the southernmost and the hottest.

Grow zones are so ingrained in US gardening culture that the hardiness zone is often noted on plant identification tags or species descriptions. In recent years, some gardening websites have begun offering search mechanisms that allow seed and plant buyers to matchmake grow zones with plants or seeds that would thrive in their area. Color-coded maps with guidance such as "Plant after March 1"—or whatever the average last frost date is in your area—continue to appear on the back of seed packets.

The descriptor "hardy to Zone 9," for example, suggests that a plant is appropriate and likely to succeed in parts of the southern United States: Texas, Louisiana, New Mexico, Arizona, Florida, and ten other states color-coded for Zone 9. Zone 9 boasts long, hot summers and mild winter temperatures that rarely fall below freezing for sustained periods, with average minimum winter temperatures dipping into the range of 20°–30° Fahrenheit.

Each zone also has two subzones, a and b, which reflect a separation of 5° Fahrenheit in minimum winter temperature.

For example:

Zone 9: Minimum average temperature 20°F–30°F

Zone 9a: Minimum average temperature 20°F–25°F

Zone 9b: Minimum average temperature 25°F–30°F.

While it's helpful to know your grow zone, more valuable, especially in a changing and erratic climate, is a basic understanding of the ecoregion in which you'll be gardening. Ecoregions take into account not only the climate and likelihood or extremity of a freeze, but also elevation, soil, natural history, and precipitation.

In San Antonio's Bexar County, for example, our hardiness zone is 9a. That means our annual average minimum temperature can hit 20°F–25°F—generally too cold for citrus trees.

However, that same sprawling area of several hundred square miles also encompasses three particular ecoregions—Blackland Prairie (known for rich, dark soils), South Texas Plains (famous for prickly brush that grows in caliche soil), and the Edwards Plateau (riddled with karst streambeds and limestone uplifts). The soil, pH levels, elevation, and precipitation in these specific ecological districts are extremely diverse and affect plant growth. Someone on the north side of town may have caliche-ridden soil, while a downtown gardener living near the San Antonio River may have clayish loamy earth.

Ecoregions are divided into 15 broad, level I ecological regions, 50 level II ecological regions that provide more detail, and 182 level III ecoregions, which are smaller ecological areas nestled within level II regions. These can be found on the EPA website. It's worth taking a moment to look up your ecoregion and grow zone so you know your specific patch of the earth.

### Annuals, perennials, biennials

Also keep in mind that plants can be annuals, perennials, or biennials.

Annuals live out their entire life cycle in one year. They die and drop their seeds, and they may return the following year if conditions are right. If too many other plants move in and crowd out the annual's progeny, or if birds eat all the seeds, or if conditions are inhospitable, the annual won't be back and you'll have to plant it again.

Perennials live multiple years. Generally the leaves, flowers, and stalks die back in the winter and then grow back from the root the following spring. Some perennials keep their foliage all year.

Biennials take two years to complete their life cycle. They might put out foliage rosettes the first year (like Mexican hat) and then flower, go to seed, and die the following year.

It's important to know whether the plants you choose are annuals, biennials, or perennials, as it makes for better planning and expectation management in the garden. Perennials are more permanent and generally cost more, but they last longer. Many gardeners use perennials as foundation plants since they can be counted on year after year. If you plant an American beautyberry bush, you know it will attain a certain height and go through its flowering and fruiting cycle at certain times, providing continuity and predictability. Annuals and biennials, while less expensive and often more showy, are more ephemeral.

Okay, so you need to find some plants. What will it be? Seeds? Plugs? Pots?

## Seeds

Seeds require more time and planning. Depending on your location or situation, you'll need to start them indoors and transplant them, or time their direct sowing into the landscape so as to avoid inhospitable conditions like frost or brutal summer heat. Most seed companies provide good guidance on a given plant's growth parameters—when to plant, how deep, how much water—so check the seed packets and follow directions for your geographic area and soil type. Seeds are definitely the most economical option but take more time.

Seeds generally sprout within a few weeks, depending on the plant. Remember that the first two green "leaves" you see on a tiny plant are not

Cotyledons are not true leaves.

True leaves.

true leaves, but food for the plant that originated in the seed. These are called the cotyledons, Latin for "seed leaf." Don't make the mistake of planting the plant prematurely when only the two cotyledons are present. Wait for the true leaves, which will follow shortly, to ensure success.

### Plugs

Plugs are seedlings, or baby plants, sometimes called "starters." You find them at nurseries and garden centers in "four-packs" or "six-packs"— small plastic containers divided into sections, each with a plant (or two) that can be transplanted into the ground and nurtured for success.

Plugs are relatively affordable and can be divided and shared with friends if you don't want all the plants for your garden. They give you a head start in the garden by providing an organism already fitted with a root system, leaves, stalks, and sometimes flowers.

### Pots

Pots are just that: vessels that contain plants for transplanting into the earth. Containers get more expensive the larger they are, ranging in size from four-inch pots to half-gallon, one-gallon, five-gallon, and bigger. While larger pots cost more, that extra investment gets you a plant that is more developed with a more advanced root system, shoots, leaves, and sometimes even flowers and fruit. Obviously, containerized plants can save you lots of growth time. Their higher cost often represents weeks, months, or even years invested by the grower in watering and care.

### How to transplant

Unless you're sowing seeds directly into the soil, at some point you'll be transplanting your plants with purpose. Every gardener has their own methodology, and here's mine.

I'm a lazy gardener by nature and prefer plants that can take care of themselves once established, so I don't use a lot of fertilizer. When I do use fertilizer, I use an organic, slow-release product in liquid or pellet form. Upon transplanting new plants, I often dip the plant in a bucket of water doused with a generous serving of slow-release liquid fertilizer until the root system is soaked before transplanting it into the ground. This can give the plant a bit of an energy boost and head start as it adjusts to its new home. Also, if more than one plant exists in a container, it makes it easier to divide the plants' root systems with the least amount of damage.

After locating the plant in the earth, I often throw a handful of slow-release fertilizer pellets on top of the soil around its base. The pellets will dissolve over time when watered.

When watering a newly transplanted plant, try to avoid blasting it with a strong stream of water. Gently watering only the soil beneath the leaves is advantageous, because if the leaves of the new transplant are wet and make contact with soil, they might develop mold or rot, which can hinder the plant's future success.

### Organic slow-release fertilizer

Organic slow-release fertilizer liquid and pellets release nutrients slowly, which is advantageous for the plant. Ideally, your soil would provide every nutrient necessary to make your plant thrive. But that's not always possible, and soil additions like slow-release fertilizer can help, especially when getting a plant started. They contain various combinations of phosphorus, nitrogen, and potassium, the primary nutrients required by plants. Think of them as plant multivitamins that are released over time, just like certain prescribed medications.

Generally, nitrogen fuels the growth of leaves, phosphorus supports the growth of flowers and fruit, and potassium helps roots grow. Such information is especially helpful if you're farming for food and need to grow corn (fruit), lettuce (leaves), or potatoes (roots). When you see "N-P-K" on fertilizer products, these are the symbols from the periodic table of the elements: N for nitrogen, P for phosphorus, and K for potassium. The higher the number for each of these, the higher the content of that nutrient. Keep in mind that plants need twenty essential elements for proper growth, but N, P, and K are the primary ones. Once again, this is not as important when choosing native plants because presumably the plants are acclimated to the ecoregion in which you are planting them. But if you are farming or growing food, it can make a big difference.

I repeat: the plants chosen here generally do not need a lot of fertilizer or other treatments. In fact, native plants sometimes react negatively to fertilizers, especially those that release all their nutrients immediately, which can "burn" the plant. That's why I treat my plants to the slow-release version as I set them up for success in the garden.

Okay, so your plant or seedling is installed in the landscape. Now what? It's not that complicated. Plants basically have four needs: air, soil, moisture, and light.

## Air

Just like us, plants breathe—but when plants do it, it's called respiration. They absorb oxygen through pores in their leaves called stomata and also through their roots. Plants also take in carbon dioxide as they grow and perform photosynthesis. In doing so, they produce oxygen.

## Soil

Knowing what kind of soil you have is helpful in choosing plants. Here are some general descriptions of soil that may prove helpful in using this book. Do the "squeeze test" by squeezing the soil in your hand to assess its status.

Clay soil is sticky and will hold its shape when squeezed. It retains moisture and nutrients but drains poorly and can hinder root growth when it's too compact or dry because less air is available for plant roots.

Sandy soil feels gritty and will fall through your fingers. It drains well but won't hold water, nutrients, or fertilizer.

Silty soil feels smooth or slippery and will hold its shape, but not as much as clay. Generally, it boasts particles that are larger than those of clay but smaller than those of sand, marrying the two profiles of sand and clay soils.

Loamy soil blends all soil textures. Loam holds its shape and falls apart when mildly poked. It's the most desirable combination of soil traits because it holds water and nutrients but allows air so as not to smother plant roots.

### What's good soil?

Few factors affect plant growth and success more than soil. Do you have a lot of critters beneath the crust—earthworms, beetles, spiders, and grubs? If so, that's a good sign that your soil is relatively healthy and contains organic matter and nutrients with a reasonable pH level.

The pH level is a way to measure the potential hydrogen in your soil. For gardening purposes, what's important to know is that the amount of hydrogen in the soil reflects a plant's ability—or inability—to absorb nutrients. The pH levels range from zero to fourteen, with zero representing superacidic soil and fourteen reflecting extremely alkaline soil. A good, healthy pH level is around six or seven.

Choosing plants that thrive in the type of soil you have in your landscape makes for an easier and more successful gardening experience. You can amend your soil by bringing in bags of "garden soil," having compost delivered, or adding organic matter like earthworm casings, but that requires more work and expense than working with what you have. You can also build soil over time, by letting leaf litter sit on the ground and self-mulch, for example. But that takes a while.

For those who want data beyond the squeeze test, consider a soil test. An easy, do-it-yourself method shared widely in gardening circles requires a soil sample, a plastic or paper cup, and vinegar.[2] Add half a cup of vinegar to the soil, and if it fizzes or bubbles, you have alkaline soil. That reaction is the response when vinegar comes into contact with acidic chemicals.

If you get no reaction, then grab another soil sample. Add half a cup of water and mix. Then add half a cup of baking soda. If it fizzes or bubbles, that indicates acidic soil—the reaction of an alkaline substance (the baking soda) coming into contact with the acids in the soil.

My soil is definitely alkaline, as bubbly as Dr Pepper.

For even more official data, contact your local agricultural cooperative extension. Most offer soil tests for free or for a small fee. You can send in a soil sample and await the analysis, which will give you facts on the type of soil you have. Soil tests are also available at nurseries and online.

### Acidic or alkaline?

Many areas of the country are known for having alkaline or acidic soils. Alkaline soils are typically loaded with lime and calcium and result in hard water. Acidic soils typically contain lots of iron and manganese. Generally, if your area has hard water, you likely have alkaline soil; areas with soft water have acidic soil.

Why does it matter? Alkaline or acidic soil will show itself in a plant's ability to absorb nutrients. That said, choosing plants native or well adapted to your area and ecosystem conditions should make this less relevant.

### Light

The vast majority of plants can grow only by conducting photosynthesis, which makes access to light critical to their health. Most of the plants mentioned here require a minimum of six hours of light per day, and not surprisingly, the best type of light for plants is natural sunlight. Sunlight provides light across the full spectrum along with all the energy plants need to grow.

Sunlight can be accessed on a windowsill, on a patio, or outside, and, if necessary, under grow lights in a greenhouse or your living room. But sunlight is definitely best, and emulating the natural dark and light cycles of the day also helps. Just like us, plants have their own circadian rhythms and darkness plays a role in their health. Leaving a plant under a bright artificial light all night is not in the plant's best interest.

When choosing a plant, assess the availability of light. Several plants with purpose work well in partial sun or partial shade. If you need to landscape in a heavily shaded area under a tree, plan accordingly.

### Water

All the plants described here do well in hot climates and generally require less than average water, once established. In the heat of a hot summer, however, even native plants wilt during peak afternoon heat. This is a plant's way of sweating, called transpiration, and allows them to conserve moisture. Adding more water won't necessarily fix the wilting, which generally abates as the sun sets.

It's better to water deeply and less often than to sprinkle lightly more frequently. The best times of day to water: early morning or early evening. Watering during the heat of the day is extremely inefficient. Much of the water is lost to evaporation and has less chance of making it to the roots where the plant can access it.

### How to tell if a plant is "established"

We often hear gardening advice along the lines of "drought tolerant, once established." What does that mean?

Being "established" for a plant means it is relatively self-sufficient and can grow, thrive, and reproduce on its own. For trees this can take years, but for perennials, biennials, and annuals, it generally takes weeks or months. And yes, natives can survive without watering. But that doesn't mean you can plunk them in the ground and walk away. You need to give them time to establish a root system and make sure they have enough moisture and nutrients to grow.

As for watering, a general rule of thumb for how often to water to get a plant "established" is the 3-2-1 rule,[3] which applies for about six weeks. The goal is to get the plant stable and self-sufficient, at which point you can revert to general watering and care for a mature organism.

Here's how the 3-2-1 rule works over six weeks:
- First three weeks—Water three times a week.
- Next two weeks—Water twice a week.
- Final week—Water just once.

Obviously this depends on the plant and the time of year. For example, most plants except perhaps succulents and cacti would not succeed if planted in August in Central Texas where temperatures are consistently in the high 90s, often hitting over 100°F. Planting and attempting to get a plant established best occurs during milder weather months so the plant is not stressed.

For already established plants, infrequent, deep watering is best. Remember to soak the root ball. I'm a hose dragger and water manually. I aim my hose with a diffuse stream of water at the bottom of the plant and count to ten. This can be very therapeutic!

# ACKNOWLEDGMENTS

This book has been a community effort, and I'm so grateful for the loads of support I've enjoyed from colleagues, friends, and family. Special thanks to my go-to plant expert, ecologist Lee Marlowe, who always shows great patience when peppered with my myriad questions. My biologist pal and native grass expert, Dr. Kelly Lyons of Trinity University, was also profoundly helpful, assisting with taxonomy and other vexing matters that reside outside my wheelhouse. Andrea DeLong Amaya and the crew at the Lady Bird Johnson Wildflower Center have always been a great resource for me and I rely heavily on their plant database. Special shout-outs to Albert del Rio and Charles Bartlett, my gardening buddies, who have provided immeasurable support over the years, and thank you, Maura Bobbitt, for coming up with the title. Texas A&M Press editor Marguerite Avery offered stellar mentorship and guidance through the sometimes challenging university press publishing process, and hats off to A&M Press's Thomas Lemmons and Emily Seyl, who first saw the potential of this project. Of course none of this would have happened without the massive support of my family, Robert, Nicolas, and Alexander Rivard, who have tolerated my many "experiments" over the years, testing recipes, gardening approaches, and theories. Wouldn't be here without you. Gracias.

# GLOSSARY

*Other useful explanations*

BUSH: woody plant usually less than three feet tall with multiple stems

CACTUS: plant with thick, fleshy stems adapted to store water, typically with thorns or spines

CULTIVAR: a cultivated variety of a plant

FULL SUN: at least six hours of sunshine a day

GRASS: plant with narrow leaves, hollow stems, and clusters of tiny, nconspicuous, usually wind-pollinated flowers

HERB: seed-bearing plant with no woody stems

HYBRID: the result of the cross-pollination of two varieties of a plant, resulting in a new plant altogether

ROOT GRAFT: a method of propagation whereby the root of one plant is grafted to the shoot of another plant to create a new variety

SHRUB (THE DRINK): a drink made with fruit, vinegar, and sugar

SHRUB (THE PLANT): see "bush"

SUCCULENT: plant with thick, fleshy stems adapted to store water

TREE: a long-lived woody plant that typically has a single primary stem with few or no branches on its lower part and upper branches that support leaves

VARIETY: a naturally occurring variation within a plant species

# RESOURCES

*Botany in a Day: The Patterns Method of Plant Identification*, by Thomas J. Elpel

Environmental Protection Agency ecoregion map, https://www.epa .gov/eco-research/ecoregions

Foraging Texas, https://www.foragingtexas.com/

Garden Style San Antonio, https://www.gardenstylesanantonio.com/

*How to Grow Native Plants of Texas and the Southwest*, by Jill Nokes

"Invasive Species Terminology: Standardizing for Stakeholder Education," Clemson University, *Journal of Extension* 58, no. 3 (June 2020), https:// tigerprints.clemson.edu/joe/vol58/iss3/27/

Lady Bird Johnson Wildflower Center, https://www.wildflower.org/

Native American Ethnobotany Database, http://naeb.brit.org/

Native Plant Society of Texas, https://npsot.org/

Texas A&M Agrilife Extension Service, https://agrilifeextension.tamu.edu/

# NOTES

## Introduction

1. Douglas W. Tallamy, *Nature's Best Hope: A New Approach to Conservation That Starts in Your Yard* (Portland: Timber Press, 2019), 9.

2. Carol Hall, "Plants That Nature Never Made," *Horticulture*, October 30, 2007, https://www.hortmag.com/plants/plants-that-nature-never-made

3. Justin Wheeler, "Picking Plants for Pollinators: The Cultivar Conundrum," Xerces Society for Invertebrate Conservation, November 21, 2017, https://xerces.org/blog/cultivar-conundrum

## Agarita

1. Elizabeth Silverthorne, *Legends of Lore of Texas Wildflowers*, 4th ed. (College Station: Texas A&M University Press, 2008) 3.

2. Patty Leslie Pasztor, phone interview with author, June 17, 2020.

3. Silverthorne, *Legends of Lore*, 3.

## American Beautyberry

1. "American Beautyberry," plant fact sheet, USDA Natural Resources Conservation Service, September 2010, https://plants.usda.gov/DocumentLibrary/factsheet/pdf/fs_caam2.pdf

2. "Learning from Our Elders: Folk Remedy Yields Mosquito-Thwarting Compound," USDA *AgResearch Magazine*, February 2006, https://agresearchmag.ars.usda.gov/2006/feb/mosquito

3. Carol Clark, "Beautyberry Leaf Extract Restores Drug's Power to Fight 'Superbug,'" Emory News Center, July 16, 2020, https://news.emory.edu/stories/2020/07/esc_beautyberry_leaf_extract/campus.html

## Bee Balm / Bergamot

1. Amy McCullough, "Hello, My Name Is Monarda," *Wildflower*, Lady Bird Johnson Wildflower Center magazine, May 27, 2018.

2. "The Buzz around Monarda," Mt. Cuba Center, February 28, 2017, https://mtcubacenter.org/buzz-around-monarda/

3. Lady Bird Johnson Wildflower Center Plant Database, accessed August 2024, https://www.wildflower.org/plants/result.php?id_plant=MODI

4. Kelly Kindscher, *Edible Wild Plants of the Prairie: An Ethnobotanical Guide* (Lawrence: University Press of Kansas, 1987), 151.

5. Jim Meuninck, *Basic Essentials: Edible Wild Plants and Useful Herbs*, Falcon Guides, 3rd ed. (Essex, CT: Falcon, 2007), 29.

### Chile Pequin or Chiltepin

1. Peppers produce the most abundant and noticeable capsaicin, but oregano, cinnamon, and cilantro also contain this chemical compound in trace amounts.

2. "Ghost Pepper-Eating Contest Leaves Man with a Hole in His Esophagus," CBS News, October 18, 2016, https://www.cbsnews.com/news/ghost-pepper-sends-man-to-hospital-hole-in-esophagus/

3. "So Hot Right Now: Why We Love the Chile Pepper," *Gastropod*, April 27, 2021, https://gastropod.com/so-hot-right-now-why-we-love-the-chile-pepper/

4. "Special Seed Starting Instructions," New Mexico State University Chile Pepper Institute, accessed August 2024, https://cpi.nmsu.edu/chile-info/growing-chile-pages/special-seed-starting-instructions.html

### Citrus: Orange, Lemon, Lime

1. Andi Wardlaw, "Rue an Overlooked Herb," Randall County Master Gardener Association, Texas A&M Horticulture Extension, accessed August 2024, https://txmg.org/randall/staying-connected/gardening-with-the-masters/gardening-tips-2/rue-an-overlooked-herb/

2. Dr. William Johnson, "Understand How Cold Temperatures Affect Citrus Trees," Texas A&M Agrilife Extension, accessed August 2024, https://aggie-horticulture.tamu.edu/newsletters/hortupdate/2011/mar/citrus_freeze.html

3. Marcus White, "James Lind: The Man Who Helped to Cure Scurvy with Lemons," BBC News, October 4, 2016, https://www.bbc.com/news/uk-england-37320399

### Cucamelon

1. Susan Mahr, "Mouse Melon or Mexican Sour Cucumber, *Melothria scabra*," Wisconsin Horticulture Division of Extension, accessed August 2024, https://hort.extension.wisc.edu/articles/mouse-melon-or-mexican-sour-cucumber-melothria-scabra/

2. Emily Rice and Kynda R. Curtis, "Drought-Tolerant Options for Southwest Agriculture: Edible Produce," Utah State University Extension, May 18, 2021, https://digitalcommons.usu.edu/extension_curall/2191/

3. Tiffany Ayuda, "Why You Should Always Eat Your Cucumbers with the Skin On," Livestrong Foundation, April 12, 2023, https://www.livestrong.com/article/467346-cucumber-peel-benefits/

4. Harni Sartika Kamaruddin, Megawati Megawati, Nurliana Nurliana, and Carla Wulandari Sabandar, "Chemical Constituents and Antioxidant Activity of *Melothria*

*scabra* Naudin Fruits," *Borneo Journal of Pharmacy* 4, no. 4 (2021), http://journal. umpalangkaraya.ac.id/index.php/bjop/article/view/2890; Anusha Govindula, Sunayana Reddy, Ponnam Manasa, and Ajay Kumar B. B., "Phytochemical Investigation and In Vitro Antidiabetic Activity of *Melothria scabra*," *Asian Journal of Pharmaceutical Research and Development* 7, no. 4 (2019): 43–48, https://www.researchgate.net /publication/337283215

## Fennel

1. Barbara Pleasant, "All about Growing Fennel," *Mother Earth News*, December 20, 2013, https://www.motherearthnews.com/organic-gardening/vegetables/growing-fennel-zw0z1312zsto

2. Marisa Jackson-Kinman, "Ayurveda and Herbs: The Amazing Fennel," California College of Ayurveda, June 30, 2017, https://www.ayurvedacollege.com/blog/ayurveda-and-herbs-amazing-fennel/

## Frogfruit

1. Jim Weber, Lynne M. Weber, and Roland H. Wauer, *Native Host Plants for Texas Butterflies: A Field Guide* (College Station: Texas A&M University Press, 2018), 95.

2. R. A. Sharma and Renu Singh, "A Review on *Phyla nodiflora* Linn.: A Wild Wetland Medicinal Herb," *International Journal of Pharmaceutical Sciences Review and Research* 20, no. 1 (May–June 2013): 57–63, http://globalresearchonline.net/journal-contents/v20-1/11.pdf

3. "7 Top Uses of Poduthalai for Hair and Skin," Wild Turmeric, June 11, 2017, https://wildturmeric.net/poduthalai-uses/

## Frostweed

1. James Richard Carter, "Flowers and Ribbons of Ice," *American Scientist* 101, no. 5 (September–October 2013): 360, https://www.americanscientist.org/article/flowers-and-ribbons-of-ice

2. T. N. Campbell, "Medicinal Plants Used by Indians," *Journal of the Washington Academy of Sciences* 41, no. 9 (September 1951): 289, https://www.biodiversitylibrary .org/page/39697188

3. Matt Warnock Turner, *Remarkable Plants of Texas: Uncommon Accounts of Our Common Natives* (Austin: University of Texas Press, 2009), 289.

4. Bob Harmes, "Observations of Crystallofolia in the 19th Century," University of Texas at Austin, Billie L. Turner Plant Resources Center, accessed August 2024, https:// w3.biosci.utexas.edu/prc/VEVI3/historical.html

## Goldenrod

1. Elizabeth Silverthorne, *Legends of Lore of Texas Wildflowers*, 4th ed. (College Station: Texas A&M University Press, 2008): 60–61.

2. Valerie Sudol, "Good as Goldenrod," Associated Press reprint in the Native Plant Society of New Jersey newsletter, Winter 2003–2004, 6–8, https://rucore.libraries .rutgers.edu/rutgers-lib/18449/PDF/1/play/

3. "Goldenrod Vinegar Recipe," Edible Wild Food, 2021, https://www.ediblewild-food.com/goldenrod-vinegar.aspx#google_vignette

**Hummingbird Mint**

1. Sushil Anand, Margaret Deighton, George Livanos, Edwin Chi Kyong Pang, and Nitin Mantri, "*Agastache* Honey Has Superior Antifungal Activity in Comparison with Important Commercial Honeys," *Scientific Reports*, December 3, 2019, https://www.ncbi.nlm.nih.gov/pmc/articles/PMC6890684/

2. Frank C. Pellett, "Anise Hyssop, Wonder Honey Plant," *American Bee Journal*, December 1940, https://ia801400.us.archive.org/26/items/CAT31355948/CAT31355948_bw.pdf

**Jimson Weed**

1. Joris J. Glas, Bernardus C. J. Schimmel, Juan M. Alba, Rocio Escobar-Bravo, Robert C. Schuurink, and Merijn R. Kant, "Plant Glandular Trichomes as Targets for Breeding or Engineering of Resistance to Herbivores," *International Journal of Molecular Sciences* 13, no. 12 (December 12, 2012): 17077–103, https://www.ncbi.nlm.nih.gov/pmc/articles/PMC3546740/

2. Amy Stewart, *Wicked Plants: The Weed That Killed Lincoln's Mother and Other Botanical Atrocities* (Chapel Hill, NC: Algonquin Books, 2009), 67.

3. Kofi Busia and Fiona Heckels, "JimsonWeed: History, Perceptions, Traditional Uses, and Potential Therapeutic Benefits of the Genus *Datura*," *HerbalGram* 69 (2006): 40–50, http://cms.herbalgram.org/herbalgram/issue69/article2930.html?ts=1595287208&signature=418e5397d443fab0ecca875cae93fc57

4. Carolyn E. Boyd and J. Philip Dering, "Medicinal and Hallucinogenic Plants Identified in the Sediments and Pictographs of the Lower Pecos, Texas Archaic," *Antiquity* 70, no. 268 (June 1996), https://go.gale.com/ps/anonymous?id=GALE%-7CA18602245&sid=googleScholar&v=2.1&it=r&linkaccess=abs&issn=0003598X-&p=AONE&sw=w

**Lantana**

1. Marilyn Sallee, "Texas Lantana," Native Plant Society of Texas newsletter, May 24, 2011, https://npsot.org/posts/texas-lantana/

2. Mark Dwyer, "2020 Is the Year of Lantana," National Garden Bureau fact sheet, 2020, https://ngb.org/wp-content/uploads/2019/10/NGB-Year-of-Lantana-Fact-sheet-printable.pdf

3. Delena Tull, *Edible and Useful Plants of the Southwest: Texas, New Mexico, and Arizona*, rev. ed. (Austin: University of Texas Press, 2013), 185.

4. Geyata Ajilvsgi, *Butterfly Gardening for Texas* (College Station: Texas A&M University Press, 2013), 197.

5. Claude Kirimuhuzya, Paul Waako, Moses Joloba, and Olwa Odyek, "The Anti-Mycobacterial Activity of *Lantana camara* a Plant Traditionally Used to Treat Symptoms of Tuberculosis in South-Western Uganda," *African Health Sciences* 9, no. 1 (March 2009): 40–45, https://www.ncbi.nlm.nih.gov/pmc/articles/PMC2932521/

6. Howard Scott Gentry, "The Warihio Indians of Sonora-Chihuahua: An Ethnographic Survey," *Bureau of American Ethnology Bulletin* 186, no. 65 (1963): 61–144.

7. Y. Rajashekar, K. V. Ravindra, and N. Bakthavatsalam, "Leaves of *Lantana camara* Linn. (Verbenaceae) as a Potential Insecticide for the Management of Three Species of Stored Grain Insect Pests," *Journal of Food Science Technology* 51, no. 11 (November 2014): 3494–99, https://www.ncbi.nlm.nih.gov/pmc/articles/PMC4571242/

## Marigold

1. "Marigold History," Burpee Seed Company, May 21, 2021, https://www.burpee.com/gardenadvicecenter/annuals/marigolds/marigold-history/article10006.html

2. Robin Lane Fox, "A Guide to the Best Exotic Marigolds," *Financial Times*, August 24, 2018, https://www.ft.com/content/e991adf4-a16a-11e8-85da-eeb7a9ce36e4

3. "How the Humble Marigold Outsmarts a Devastating Tomato Pest," *Science Daily*, Newcastle University, March 2019, https://www.sciencedaily.com/releases/2019/03/190301160909.htm

4. Robert Trostle Nehrer, "The Ethnobotany of *Tagetes*," *Economic Botany* 22, no. 4 (October–December 1968): 317–25, https://www.jstor.org/stable/4252990?read-now=1&seq=1#page_scan_tab_contents

## Mexican Hat

1. Gabriel Ruben Bernier, "Ethnobotany of the Northern Cheyenne: Medicinal Plants," University of Montana, 2004, https://scholarworks.umt.edu/cgi/viewcontent.cgi?article=10335&context=etd

## Milkweed

1. Anurag Agrawal, *Milkweed and Monarchs: A Migrating Butterfly, a Poisonous Plant, and Their Remarkable Story of Coevolution* (Princeton, NJ: Princeton University Press, 2017), 152.

2. "Monarch Disease," Frequently Asked Questions, Monarch Joint Venture, accessed August 2024, https://monarchjointventure.org/faq/monarch-disease

3. Patricia Cox Crews, Shiela A. Sievert, Lisa T. Woeppel, and Elizabeth A. Mccullough, "Evaluation of Milkweed Floss as an Insulative Fill Material," *Textile Research Journal* 61, no. 4 (April 1991), https://journals.sagepub.com/doi/10.1177/004051759106100403

## Oaks

1. Henry David Thoreau, *The Writings of Henry David Thoreau*, vol. 7 (Boston: HoughtonMifflin, 1892), 84, https://en.wikisource.org/wiki/Page%3AAutumn._From_the_Journal_of_Henry_D._Thoreau.djvu/98

2. Marcie Lee Mayer, *Eating Acorns: Field Guide—Cookbook—Inspiration* (self-pub., April 2017).

3. Dasl Yoon and Na-Young Kim, "Humans Are Gobbling Up Acorns, Driving Squirrels Nuts," *Wall Street Journal*, October 23, 2019, https://www.wsj.com/articles/humans-are-gobbling-up-acorns-driving-squirrels-nuts-11571840532?mod=searchresults&page=1&pos=1

4. Li Liu, Sheahan Bestel, Jinming Shi, Yanhua Song, and Xingcan Chen, "Paleolithic Human Exploitation of Plant Foods during the Last Glacial Maximum in North China," *Proceedings of the National Academy of Sciences* 110, no. 14 (February 11, 2013): 5380–85, https://www.pnas.org/doi/10.1073/pnas.1217864110

5. Mary Lynn Ritzenthaler and Catherine Nicholson, "The Declaration of Independence and the Hand of Time," *Prologue* 48, no. 3 (Fall 2016), https://www.archives.gov/publications/prologue/2016/fall/declaration

## Passionflower

1. John L. Neff, "The Passionflower Bee: *Anthemurgus passiflorae*," *Passiflora: The Journal & Newsletter of Passiflora Society International* 13, no. 1 (Spring/Summer 2003), http://www.sbs.utexas.edu/philjs/pdf/TheBee.pdf

2. T. R. Radhamani, L. Sudarshana, and Rani Krishnan, "Defense and Carnivory: Dual Role of Bracts in *Passiflora foetida*," *Journal of Biosciences* 20 (1995): 657–64, https://link.springer.com/article/10.1007/BF02703305

## Peppergrass

1. Leda Meredith, "Foraging Wild Peppergrass for a Native Spice," *Mother Earth News*, June 19, 2020, https://www.motherearthnews.com/real-food/foraging-wild-peppergrass-for-a-native-spice-zbcz1407/

## Prickly Pear

1. Rod Santa, "A Prickly Subject: South Texas Ranchers Turn to Cactus to Feed Cattle," *Bryan–College Station (TX) Eagle*, updated July 21, 2020, https://theeagle.com/landandlivestockpost/a-prickly-subject-south-texas-ranchers-turn-to-cactus-to-feed-cattle/article_c47fe68b-1a6f-51e6-8701-7c2aaf7b173a.html

2. Carl Olson, *50 Common Insects of the Southwest* (Tucson: Western National Parks Association, 2004), 8.

3. Bastiaan M. Drees and John A. Jackman, *A Field Guide to Texas Common Insects* (Houston: Gulf Publishing, 1998), 83.

4. Michael Zeece, *Introduction to the Chemistry of Food* (Amsterdam: Academic Press, 2020), 313–44.

5. Charles W. Kane, *Medicinal Plants of the American Southwest* (Lincoln City, OR: Lincoln Town Press, 2017), 199.

6. Audrey L. Buttice, Joyce M. Stroot, Daniel V. Lim, Peter G. Stroot, and Norma A. Alcantar, "Removal of Sediment and Bacteria from Water Using Green Chemistry," *Environmental Science & Technology* 44, no. 9 (2010): 3514–19, https://pubs.acs.org/action/showCitFormats?doi=10.1021%2Fes9030744&href= doi/10.1021%2Fes9030744

## Red Yucca

1. B. L. Turner and Matt W. Turner, "Natural Populations of *Hesperaloe* (Agavaceae) in Texas," *Lundellia*, no. 5 (December 2002), https://w3.biosci.utexas.edu/prc/pdfs/Turner_Lundellia05.pdf

2. Jacqueline Soule, "Edible Flowers from Desert Landscapes," Gardening with Soule, March 20, 2022, https://gardeningwithsoule.wordpress.com/2022/03/08/edible-flowers-from-desert-landscapes/

3. Delena Tull, *Edible and Useful Plants of Texas and the Southwest: A Practical Guide* (Austin: University of Texas Press, 1999), 379.

## Sage

1. Biljana Bauer Petrovska, "Historical Review of Medicinal Plants' Usage," *Pharmacognosy Reviews* 6, no. 11 (January–June 2012): 1–5, https://www.ncbi.nlm.nih.gov/pmc/articles/PMC3358962/#ref2

2. Frank Chapman Pellett, *American Honey Plants* (Hamilton, IL: American Bee Journal, 1920), 225, https://horizontalhive.com/download-free/american-honey-plants-pellett.pdf

3. Adrian L. Lopresti, "Salvia (Sage): A Review of Its Potential Cognitive-Enhancing and Protective Effects," *Drugs in R&D* 17, no. 1 (March 2017): 53–64, https://www.ncbi.nlm.nih.gov/pmc/articles/PMC5318325/

4. Magda Feres, Luciene C. Figueiredo, Ilizvania M. Q. Barreto, Mary Hellen M. Coelho, Marcelo W. B. Araujo, and Sheila C. Cortelli, "In Vitro Antimicrobial Activity of Plant Extracts and Propolis in Saliva Samples of Healthy and Periodontally-Involved Subjects," *Journal of the International Academy of Periodontology* 7, no. 3 (July 2005): 90–96, https://pubmed.ncbi.nlm.nih.gov/16022025/

5. Petrovska, "Historical Review."

## Sunflower

1. Christy Bassett, "Saving and Eating Sunflower Seeds," Northeast Organic Farming Association, August 20, 2021, https://www.nofamass.org/articles/2021/08/saving-and-eating-sunflower-seeds/

2. Kelly Kindscher, *Edible Wild Plants of the Prairie: An Ethnobotanical Guide* (Lawrence: University Press of Kansas, 1987), 127.

3. Spherical Insights, "Global Sunflower Oil Market Size Worth USD 30.52 Billion by 2032 | CAGR of 6.72%," May 31, 2023, https://www.sphericalinsights.com/press-release/sunflower-oil-market

## Turk's Cap

1. Geyata Ajilvsgi, *Butterfly Gardening for Texas* (College Station: Texas A&M University Press, 2013), 154.

## Wild Onion

1. Richard S. Rivlin, "Historical Perspective on the Use of Garlic," *Journal of Nutrition* 131, no. 3 (March 2001): 951S–954S, https://academic.oup.com/jn/article/131/3/951S/4687053

## From the Ground Up

1. The first version of a hardiness zone map, which later fell out of use, was published in Alfred Rehder's *Manual of Cultivated Trees and Shrubs* in 1927, https://archive.org/details/manualofcultivat0000rehd/mode/2up

2. "Garden Soil Testing," Clapp Library, Wellesley College handout on at-home soil testing, https://www.clapplibrary.org/wp-content/uploads/2020/04/SoilSample.pdf

3. Mark Peterson, "Countdown to Establishing New Plants: 3-2-1," Garden Style San Antonio, January 9, 2014, https://www.gardenstylesanantonio.com/garden-tips-blog/2014-01-countdown-to-establishing-new-plants-3-2-1/

# INDEX

Note: Italicized page numbers indicate material in photographs or illustrations.